Birkhäuser

Compact Textbooks in Mathematics

This textbook series presents concise introductions to current topics in mathematics and mainly addresses advanced undergraduates and master students. The concept is to offer small books covering subject matter equivalent to 2- or 3-hour lectures or seminars which are also suitable for self-study. The books provide students and teachers with new perspectives and novel approaches. They may feature examples and exercises to illustrate key concepts and applications of the theoretical contents. The series also includes textbooks specifically speaking to the needs of students from other disciplines such as physics, computer science, engineering, life sciences, finance.

- **compact:** small books presenting the relevant knowledge
- **learning made easy:** examples and exercises illustrate the application of the contents
- **useful for lecturers:** each title can serve as basis and guideline for a semester course/lecture/seminar of 2-3 hours per week.

Kiyan Naderi • Konstantin Pankrashkin

Introduction to Spectral Graph Theory

Birkhäuser

Kiyan Naderi
Institut für Mathematik
Universität Oldenburg
Oldenburg, Niedersachsen, Germany

Konstantin Pankrashkin
Institut für Mathematik
Universität Oldenburg
Oldenburg, Niedersachsen, Germany

ISSN 2296-4568   ISSN 2296-455X   (electronic)
Compact Textbooks in Mathematics
ISBN 978-3-032-01707-9   ISBN 978-3-032-01708-6   (eBook)
https://doi.org/10.1007/978-3-032-01708-6

This book is a translation of the original German edition "Einführung in die spektrale Graphentheorie" by Kiyan Naderi and Konstantin Pankrashkin, published by Springer Nature Switzerland AG in 2024. The translation was done with the help of an artificial intelligence machine translation tool. A subsequent human revision was done primarily in terms of content, so that the book will read stylistically differently from a conventional translation. Springer Nature works continuously to further the development of tools for the production of books and on the related technologies to support the authors.

Translation from the German language edition: "Einführung in die spektrale Graphentheorie" by Kiyan Naderi and Konstantin Pankrashkin, © The Editor(s) (if applicable) and The Author(s), under exclusive licence to Springer Nature Switzerland AG 2024. Published by Springer Nature Switzerland. All Rights Reserved.

This book is published under the imprint Birkhäuser, www.birkhauser-science.com by the registered company Springer Nature Switzerland AG
The registered company address is: Gewerbestrasse 11, 6330 Cham, Switzerland

If disposing of this product, please recycle the paper.

Preface to the English Edition

In spectral graph theory, various properties of graphs are investigated using the methods of linear algebra, in particular by means of the eigenvalues and eigenvectors of various matrices that describe the graph structure. Interest in various aspects of graph theory is apparently increasing, as many notions and results are applied in the context of data science. The calculation of the most important invariants for large graphs is an NP-complete problem, whereas the eigenvalues and eigenvectors can be found in polynomial time and provide non-trivial estimates for the graph invariants. Furthermore, the eigenvectors of various graph matrices are also one of the most important tools in the cluster analysis of networks [110]. Other important advances in computer science are also closely linked to the spectral theory of graphs [37,67]. Although the central topics are covered by the well-known monographs of Bapat [11], Brouwer, Haemers [24], Colin de Verdière [31], Chung [27], Cvetković, Doob, Sachs [38], Cvetković, Rowlinson, Simić [41], van Mieghem [81] and are also represented in some books on general graph theory, e.g., Bollobás [21] and Godsil, Royle [56], we always had the feeling that there are hardly any standard works in this field that could be used directly as a textbook for a relevant mathematical course with clearly defined prerequisites. This impression is particularly reinforced by very uneven literature recommendations for courses on the subject at various universities. Probably for this very reason, spectral graph theory has so far been regularly represented in very few mathematics degree programs. With this book, we present our attempt to write a text that could meet these open needs. We were conceptually inspired by the texts of Nica [87], Spielman [98], and Williamson [116].

The selection of topics in the book essentially corresponds to the personal curiosity and scientific background of the authors as well as the interests of their immediate environment: It must be taken into account that the two authors of the book are mainly concerned with the spectral theory of differential operators and have entered a new field with this text. An attempt was made to address a wide range of questions, the motivation for which remains understandable without particularly deep knowledge of graph theory. The treatment of the questions is primarily designed as an intermediate step toward the analysis of differential operators. All topics can be dealt with using elementary methods from traditional basic mathematical lectures, but at many points in the book, various parallels to corresponding advanced results of "continuous" geometric analysis are also discussed.

In Chap. 1, basic concepts are introduced and the main objects of spectral graph theory (adjacency matrix and Laplace matrix) are presented. Then important facts of linear algebra about characteristic polynomials, eigenvalues, and eigenvectors of matrices are repeated, and the spectral theorem for self-adjoint linear mappings is proven. With these elementary tools, first spectral results about graph matrices are then derived. In Chap. 2 we prove the min-max formulas for eigenvalues as well as the Perron-Frobenius theorem. With this, several standard results of classical spectral theory of graphs are proven (characterization of bipartite graphs, inequalities between eigenvalues and graph invariants, etc.). Although there are various proofs for many of these results, we consciously decided on a uniform approach via the min-max principle. From Chap. 3 onwards, weighted Laplace matrices are also considered. The focus is no longer on eigenvalues but on eigenfunctions, in particular we prove the Cheeger inequality and the nodal theorem for eigenfunctions. Some of the results presented in Chap. 3 have only been known for about 10 years and are, to our knowledge, presented for the first time in a textbook. In Chap. 4, the classical theorems of Kuratowski and Wagner on planar graphs are proven. Then planarity is also characterized by the invariant of Colin de Verdière (where minor monotonicity is initially assumed without proof). In the last three Sects. 4.6–4.8, we give a complete proof for the minor monotonicity: To our knowledge, these topics have also not yet been treated in textbook format.

We would like to emphasize that the book is only intended as a thorough introduction that can be worked through within a semester, and that it does not claim to provide a complete representation of the entire research area. We hope that with the knowledge acquired, one will be able to read more advanced texts independently: For deepening on individual topics, appropriate literature recommendations are given in the text. Spectral graph theory is a relatively young field of research: Often the review article published by Collatz and Sinogowitz in 1957 [33] is considered the official foundation.[1] Therefore, we have tried to cite the original works for most statements. Some exercises also come directly from current research articles. With this, we want to explicitly emphasize that many of the important results were only proven in the last decades. We have assigned the results to individual persons to the best of our knowledge and belief: We do not want to take any responsibility for this and would be very thankful for additional historical comments. The content of the book corresponds to the content of the course "Special Topics in Analysis: Analysis on Graphs", which was offered at the Institute for Mathematics of the Carl von Ossietzky University of Oldenburg in the format 3+1 SWS in the years 2021–2023. In other words, the essential learning material with associated exercises could be covered in 28 classes of 90 minutes each (21 lectures and 7 exercise sessions). The text is self-contained and all necessary terms and theorems of graph theory, linear algebra, and analysis are fully proven or repeated. The aforementioned course was offered as part of the master's program with a focus on analysis, but all topics should also be accessible to bachelor's students. The topics of the first three chapters only

---

[1] There were also earlier works: A comprehensive literature review can be found in the book by Cvetković et al. [38].

Fig. 1 Content dependencies
between the book sections

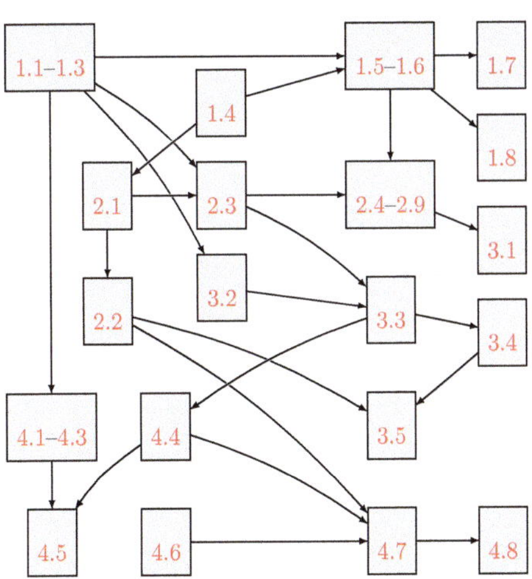

require basic knowledge of linear algebra and analysis, which is usually acquired in the first two semesters of a bachelor's degree in mathematics. More specifically, we believe that understanding the material of Sect. 1.4 is sufficient to meaningfully work on all topics of Chaps. 1–3.

Implicitly, we hope that in this new framework the learning material of the basic lectures will also appear more attractive. For the investigation of Colin de Verdière's invariant in Chap. 4, some basics on submanifolds are needed, which are also covered in the same chapter. Therefore, the book is also suitable for a reading course on linear algebra or analysis. The approximately 50 exercises included in the book were sufficient for our semester course. Important exercises that are used in the main text are underlined. Further exercises can be found in the previously cited monographs, especially in Bollobás [21, Ch. 8], Godsil, Royle [56, Ch. 8–13], and Stanić [100]. The topics can simply be worked on in the suggested order, but other variants are also possible: To enable a rescheduling, the dependencies between different topics are shown in Fig. 1.

For helpful discussions and enlightening comments on preliminary versions of the German edition, we would like to thank Ram Band, Gregory Berkolaiko, Daniel Grieser, Matthias Keller, Delio Mugnolo, Olaf Post, and Stefan Steinerberger. We are also very grateful to Phil Bergwitz for the detailed proofreading of individual sections.

The German edition was published in November 2024, and, to our knowledge, it was already used for classes at the universities of Graz, Hagen, Potsdam, and Trier. Some (minor) errors and typos were corrected in this English version, but no further changes in the contents were made. We hope that this edition will reach a wide readership!

Oldenburg, Germany   Kiyan Naderi
June 30, 2025   Konstantin Pankrashkin

Declarations

Competing Interests The authors have no competing interests to declare that are relevant to the content of this manuscript.

Contents

Elementary Theory

In this chapter, we introduce the basic concepts and main motivations of spectral graph theory, which will be actively used throughout the rest of the book. We also revisit some topics from linear algebra. With their help, we derive classical results about the spectra of the most important graph matrices.

1.1   Basic Concepts of Graph Theory

We will first discuss the most important concepts and results of graph theory. Our discussion is kept very minimalist, but it is sufficient to handle the contents of spectral graph theory presented in this text. A more detailed presentation can be found, for example, in the books by Aigner [2], Biggs [18], Bretto et al. [22], Diestel [45], Volkmann [109] or West [113].

For a set Z, the number of elements in Z is denoted by $|Z|$ or $\#Z$.

Definition 1.1.1 A **graph** is a pair $G := (X, E)$ consisting of a set X and a set E, such that each element of E is a two-element subset of X.

- The elements of X are called **vertices** (or **nodes**) of G and the elements of E are called **edges** of G.
- If $x, y \in X$, with $\{x, y\} \in E$, then we write $x \sim y$ and say that the vertices x and y are **adjacent** or **neighboring**.
- If $x \in X$ and $e \in E$ with $x \in e$, then we write $x \sim e$ or $e \sim x$ and say that the vertex x and the edge e are **incident**.
- Two edges $e, e' \in E$ with $e \neq e'$ are **neighboring**, if $e \cap e' \neq \varnothing$, i.e., if e and e' share a common vertex. In this case, we also write $e \sim e'$.

The objects defined above are illustrated in Fig. 1.1.

© The Author(s), under exclusive license to Springer Nature Switzerland AG 2025
K. Naderi, K. Pankrashkin, *Introduction to Spectral Graph Theory*, Compact
Textbooks in Mathematics, https://doi.org/10.1007/978-3-032-01708-6_1

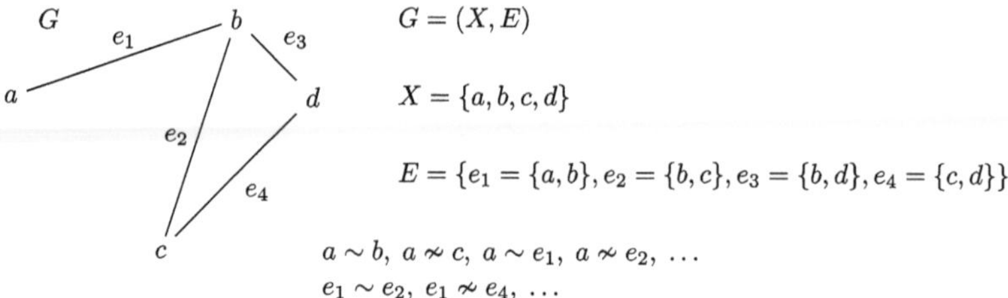

Fig. 1.1 Edges and vertices of an example graph

Fig. 1.2 Example of an orientation

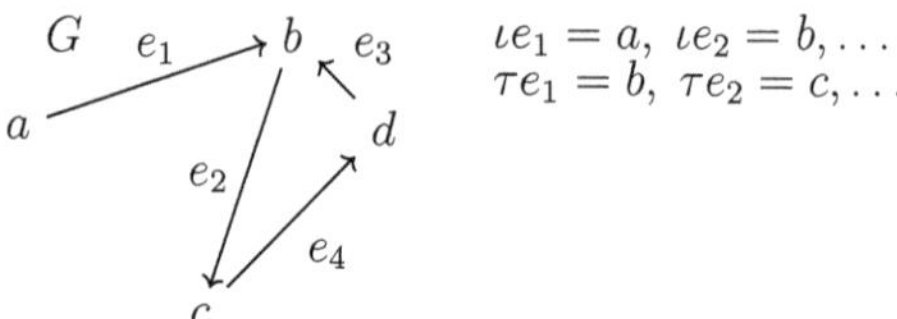

Remark 1.1.2

- In this book, from now on, we will always understand a graph to be a finite non-empty graph, i.e. $G = (X, E)$ **with a finite and non-empty set** X (the set of edges E is allowed to be empty).
- The definition of the set of edges does not a priori allow multiple edges (i.e., several edges between two vertices) or loops (edges from a vertex to itself): from the two-element nature of the subsets for the edges, it follows from $\{x, y\} \in E$ automatically $x \neq y$.
- It is also possible to consider "generalized edges" that connect more than two vertices: The corresponding combinatorial structures are called *hypergraphs*. Such objects are not studied in this book, but we refer to the article by Mulas et al. [86] for a compact introduction and further references.

The edges of a graph are "symmetric" by definition (have no traversal direction), but it is often convenient to introduce a traversal direction, as this makes some formulas much more compact to write.

Definition 1.1.3 Let $G := (X, E)$ be a graph. An **orientation** on G is a mapping $\iota : E \to X$ with $\iota e \in e$ for all $e \in E$. For each edge $e \in E$, the vertex ιe is interpreted as the **starting vertex** of e. Then there is a uniquely determined vertex τe with $e = \{\iota e, \tau e\}$: The vertex τe is referred to as the **end vertex** of e (see Fig. 1.2). Obviously, there are $2^{|E|}$ different orientations on G, as there are two possibilities for ιe for each $e \in E$.

Let $G := (X, E)$ be a graph. Define the **degree mapping** deg:

$$\deg : \ X \to \mathbb{N}_0, \qquad \deg x := \#\{e \in E : e \sim x \text{ in } G\} \equiv \#\{y \in X : \{x, y\} \in E\}.$$

When working with multiple graphs at the same time, we also write $\deg_G$ instead of deg. A vertex $x \in X$ is called:

- **isolated** in G, if $\deg x = 0$,
- **pendant** in G, if $\deg x = 1$.

The graph G is called

- **complete**, if $x \sim y$ for all $x, y \in X$ with $x \neq y$ (i.e., if any two vertices are connected by an edge: This is obviously equivalent to $\deg x = |X| - 1$ for all $x \in X$),
- **empty**, if $E = \varnothing$ (i.e., if G has no edges and thus consists only of isolated vertices).

By induction on $|X|$ or $|E|$, one directly proves:

Lemma 1.1.4 (Handshake Lemma) *In every graph $G = (X, E)$, it holds that*

$$\sum_{x \in X} \deg x = 2|E|.$$

The structure of a graph G can be described using various matrices.

- The **adjacency matrix** of G is

$$A := (A_{x,y})_{x,y \in X}, \qquad A_{x,y} := \begin{cases} 1, & x \sim y, \\ 0, & \text{otherwise.} \end{cases}$$

From A, one can derive other important matrices. Particularly often used are:

- the **degree matrix**

$$D := (D_{x,y})_{x,y \in X}, \qquad D_{x,y} := \begin{cases} \deg x, & x = y, \\ 0, & \text{otherwise,} \end{cases}$$

- the **Laplace matrix** $L := D - A$,
- the **signless Laplace matrix** $Q := D + A$.

Example 1.1.5 For the graph in Fig. 1.3, we have:

Fig. 1.3 Example graph

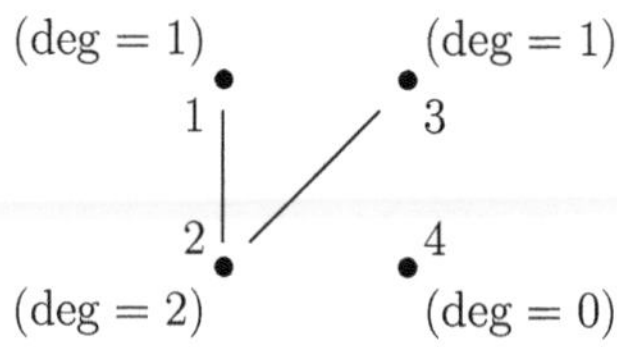

$$A = \begin{pmatrix} 0 & 1 & 0 & 0 \\ 1 & 0 & 1 & 0 \\ 0 & 1 & 0 & 0 \\ 0 & 0 & 0 & 0 \end{pmatrix}, \qquad D = \begin{pmatrix} 1 & 0 & 0 & 0 \\ 0 & 2 & 0 & 0 \\ 0 & 0 & 1 & 0 \\ 0 & 0 & 0 & 0 \end{pmatrix},$$

$$L = \begin{pmatrix} 1 & -1 & 0 & 0 \\ -1 & 2 & -1 & 0 \\ 0 & -1 & 1 & 0 \\ 0 & 0 & 0 & 0 \end{pmatrix}, \qquad Q = \begin{pmatrix} 1 & 1 & 0 & 0 \\ 1 & 2 & 1 & 0 \\ 0 & 1 & 1 & 0 \\ 0 & 0 & 0 & 0 \end{pmatrix}.$$

Remark 1.1.6 We note that A, D, L, Q are always square symmetric $X \times X$ matrices and D is even a diagonal matrix. All these matrices, as well as other $X \times X$ matrices that are introduced later and take into account the structure of G in a suitable way will be called **graph matrices** of G. To emphasize the dependence on G, A_G is often written instead of A, L_G instead of L, etc. In this book, we deal with the question of how certain structural properties of G can be read from the associated graph matrices, especially from their eigenvalues and eigenfunctions. In the first Sects. 1.1–1.3, we first describe various families of properties that a graph can in principle possess.

Remark 1.1.7 (Possible Generalizations of Graph Matrices) In the definition of the adjacency matrix (and the derived matrices), it is explicitly assumed that all edges have the same "weight", so only the presence or absence of an edge between two vertices is marked (with 1 or 0). The construction can be generalized by assigning any *positive* weight to each edge: This results in weighted graph matrices, which we examine in later chapters (see Sect. 3.3).

It is also possible to assign negative weights to some edges. This can model many situations: For example, a relationship between two people can be a friendship or an enmity, or one can think of likes/dislikes of publications in social networks. This leads to the concept of a signed graph: This class is not covered in our book, but we refer to the articles [13, 71, 101] and the further works cited there for a good introduction.

It is possible to consider oriented adjacency matrices: i.e., $A_{x,y} = 1$ exactly when there is an edge e with $\iota e = x$ and $\tau e = y$. In this case, the matrix A (as well as further matrices derived from A) is not necessarily symmetric, which greatly complicates the analysis. Some questions for this new framework are discussed, for example, in the book by Brouwer and Haemers [24].

In many applications, infinite graphs (also with multiple edges) are considered: In this case, most definitions and the notation must be adapted. The associated infinite graph matrices are considered as linear operators in suitable Hilbert or Banach spaces and studied with methods of functional analysis, see e.g. the monographs by Grigor'yan [60], Keller et al. [69], Kostenko and Nicolussi [70], Mugnolo [85], Post [92], Soardi [97], Woess [118], as well as the review article by Mohar and Woess [83].

Graphs are usually considered "up to numbering of the vertices":

Definition 1.1.8 Two graphs $G := (X, E)$ and $G' := (X', E')$ are called **isomorphic** if G' arises from G by "renumbering the vertices", i.e., if there exists a bijective mapping $\varphi : X \to X'$ such that for any $x, y \in X$ it holds that

$$\{x, y\} \in E \text{ if and only if } \{\varphi(x), \varphi(y)\} \in E'.$$

The relation of isomorphism is obviously an equivalence relation on the set of graphs, and a "graph property" is understood to be a family of graphs that is closed under isomorphisms. Of particular interest are the so-called **graph invariants** ("numerical properties") that assign a number to each graph, but only depend on the associated equivalence class. The simplest graph invariants include, among others,

$$G = (X, E) \mapsto |X| \text{ (number of vertices in } G),$$

$$G = (X, E) \mapsto |E| \text{ (number of edges in } G),$$

$$G = (X, E) \mapsto \deg_{\max} G := \max_{x \in X} \deg x \text{ (maximum degree in } G)$$

$$\text{(often also denoted as } \Delta(G)),$$

$$G = (X, E) \mapsto \deg_{\min} G := \min_{x \in X} \deg x \text{ (minimum degree in } G)$$

$$\text{(often also denoted as } \delta(G)),$$

$$G = (X, E) \mapsto \deg_{\mathrm{avg}} G := \frac{1}{|X|} \sum_{x \in X} \deg x \text{ (average degree in } G),$$

$$G = (X, E) \mapsto \beta(G) := |E| - |X| + 1 \text{ (the first Betti number)}.$$

Further invariants are introduced in Sect. 1.3 and later in the text.

Definition 1.1.9 (Subgraphs) Let $G := (X, E)$ be a graph.

- A graph $G' := (X', E')$ is a **subgraph** of G if $X' \subset X$ and $E' \subset E$; in this case, we write $G' \subset G$.

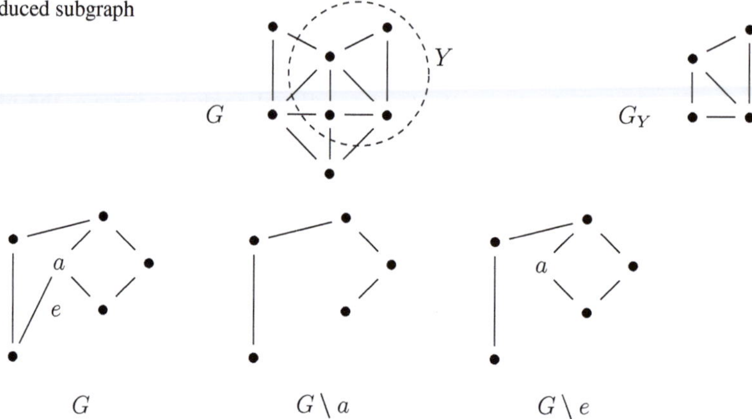

Fig. 1.4 Induced subgraph

Fig. 1.5 Removal of vertices and edges

- Let $Y \subset X$. The **subgraph** of G **induced by** Y is $G_Y := (Y, E_Y)$ with

$$E_Y := \big\{ \{y, y'\} \in E : y, y' \in Y \big\},$$

see Fig. 1.4.

Definition 1.1.10 (Operations on Graphs) Let $G := (X, E)$ be a graph.

- Let $E_0 \subset E$, then $G \setminus E_0$ denotes the graph obtained by removing all edges $e \in E_0$, i.e., $G \setminus E_0 := (X, E')$ with $E' := E \setminus E_0$. If E_0 consists of only one edge, $E_0 := \{e\}$, then we often write $G \setminus e$ instead of $G \setminus \{e\}$.
- Let $X_0 \subsetneq X$, then $G \setminus X_0$ denotes the graph obtained by removing all vertices $x \in X_0$ and all edges incident to these vertices, i.e., $G \setminus X_0 := (X', E')$ with

$$X' := X \setminus X_0, \quad E' := E \setminus \{e \in E : \text{There exists } x \in X_0 \text{ with } e \sim x\}.$$

If X_0 consists of a single vertex, $X_0 := \{x\}$, then we often write $G \setminus x$ instead of $G \setminus \{x\}$.

The constructions are illustrated in Fig. 1.5.

Further graph operations are considered in Definition 1.1.25.

Definition 1.1.11 Let $G := (X, E)$ be a graph.

- The **complement graph** of G is the graph $\overline{G} := (X, \overline{E})$ with (see Fig. 1.6)

$$x \sim y \text{ in } \overline{G} \text{ if and only if } \{x, y\} \notin E.$$

Fig. 1.6 Complement graph

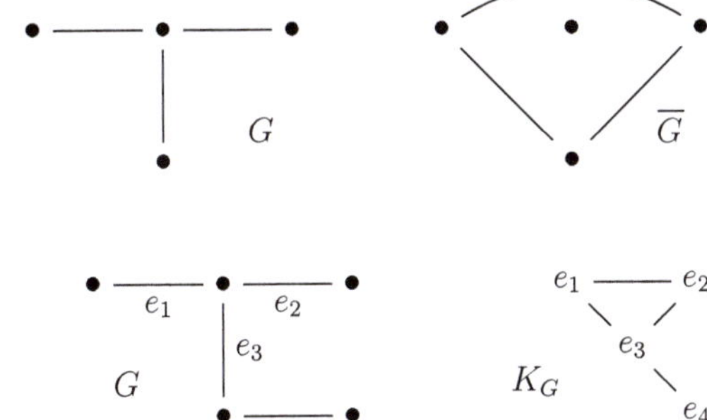

Fig. 1.7 Line graph

- The **line graph** of G is the graph $K_G := (X', E')$,

$$X' := E, \quad \{e, e'\} \in E' \text{ if and only if } e \sim e' \text{ in } G,$$

see Fig. 1.7.

Definition 1.1.12 Let $G := (X, E)$ be a graph. Let $x, y \in X$ and $k \in \mathbb{N}$.

- A **walk** of length k from x to y in G is a sequence

$$(x_0, x_1, \ldots, x_k) \subset X$$

such that $x_0 = x$, $x_k = y$ and $x_{j-1} \sim x_j$ for all $j \in \{1, \ldots, k\}$. In this case, we also say that the walk connects the vertices x and y. For $x = y$, the walk is called **closed**.
- If all $x_0, \ldots, x_k$ are distinct, the walk is called a **path** of length k.
- If $x = y$, $k \geq 3$, and $x_0, \ldots, x_{k-1}$ are distinct, the walk is called a **closed path** or a **cycle** of length k.

See Fig. 1.8.

Definition 1.1.13 A graph $G := (X, E)$ is called **connected**, if any $x, y \in X$ with $x \neq y$ are connected by a path. The **distance** $d(x, y)$ between x and y (with $x \neq y$) is the length of the shortest path between x and y. For $x = y$ we set $d(x, y) := 0$.

As an exercise, one can show:

Proposition 1.1.14 *Let $G := (X, E)$ be a graph and let $x, y \in X$ with $x \neq y$. If there exists a walk from x to y, then there also exists a path from x to y. In particular, the following conditions are equivalent:*

(a) G is connected,
(b) arbitrary $x, y \in X$ with $x \neq y$ are connected by a walk.

Fig. 1.8 Walk, path, and cycle

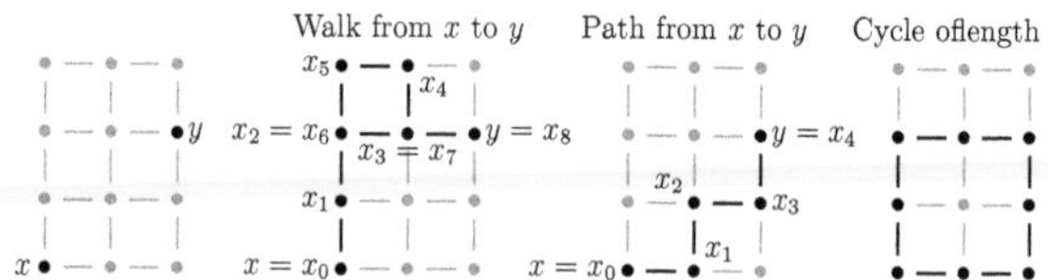

Fig. 1.9 Connected components

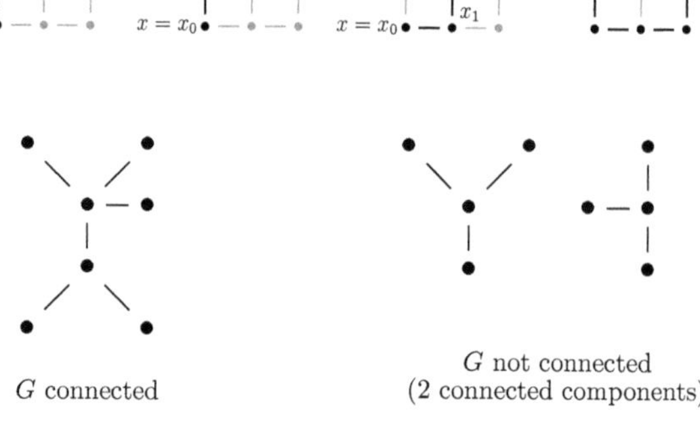

G connected

G not connected
(2 connected components)

Definition 1.1.15 Let $G := (X, E)$ be a graph.

- A subgraph $G' \subset G$ is a **connected component** of G, if:
  - G' is connected <u>and</u>
  - G' is not contained in any other connected subgraph of G (see Fig. 1.9).
- If $Y \subset X$, such that the induced graph G_Y is connected, then Y is called **connected in G**.

Remark 1.1.16 If $G = (X, E)$ and $G_i = (X_i, E_i)$ with $i \in \{1, \ldots, p\}$ are the connected components of G, with $X = \bigcup_{i=1}^{p} X_i$, then we write

$$G = G_1 \sqcup \cdots \sqcup G_p.$$

If A_i are the adjacency matrices of G_i, then the adjacency matrix A of G has the block representation (with appropriate numbering)

$$A = \begin{pmatrix} A_1 & 0 & 0 & \ldots \\ 0 & A_2 & 0 & \ldots \\ 0 & 0 & A_3 & \ldots \\ \vdots & \vdots & \vdots & \ddots \end{pmatrix},$$

similar representations also apply for the Laplace matrices L and L_i of G and G_i etc.

Definition 1.1.17 A **forest** is a graph without cycles (i.e., a graph such that no subgraph of it is a cycle). A **tree** is a <u>connected</u> graph without cycles (Fig. 1.10).

Fig. 1.10 Trees

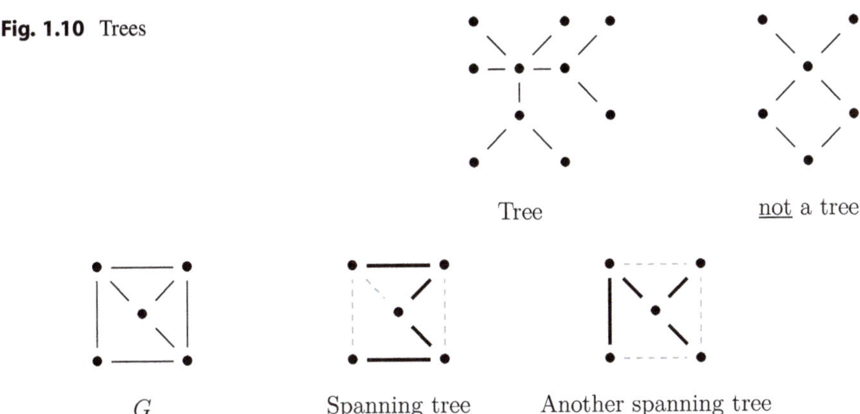

Tree   <u>not</u> a tree

G Spanning tree   Another spanning tree

Fig. 1.11 Spanning trees

Remark 1.1.18 Every connected component of a forest is obviously a tree. That is, every forest consists of trees (hence the name "forest").

Trees form a very important class of graphs and they can be characterized by various means. The following statements are proven as an exercise (see Exercise 1.1.1 and 1.1.3):

Proposition 1.1.19 *Let $G := (X, E)$ be a graph, then the following conditions are equivalent:*

- *G is a tree,*
- *for arbitrary $x, y \in X$ with $x \neq y$ there is <u>exactly one</u> path from x to y.*
- *G is connected with $|X| = |E| + 1$.*
- *G contains no cycles and $|X| = |E| + 1$.*

Definition 1.1.20 Let $G := (X, E)$ be a graph. A **spanning tree** of G is a subgraph of G that is a tree and contains all vertices of G (Fig. 1.11).

As an exercise, one shows (Exercise 1.1.2):

Proposition 1.1.21 *Every connected graph has a spanning tree.*

Corollary 1.1.22 *For every connected graph $G = (X, E)$ the inequality $|E| \geq |X| - 1$ holds.*

Proof Let $G' = (X, E')$ be a spanning tree in G, then $|E'| = |X| - 1$ according to Proposition 1.1.19 and $|E| \geq |E'|$ thanks to $E' \subset E$. $\qquad\square$

Bipartite graphs form another important class.

Definition 1.1.23 A graph $G := (X, E)$ is called **bipartite**, if there is a decomposition $X = Y \cup Y'$ into disjoint subsets Y and Y' such that each edge connects a vertex from Y with a vertex in Y'. The sets Y and Y' are called **partition components**.

It should be noted that one of the sets Y, Y' in the definition may also be empty. In particular, all empty graphs are bipartite. As an exercise, one shows that every tree is also a bipartite graph.

As already seen above, graphs $G := (X, E)$ are often graphically represented: Each vertex $x \in X$ corresponds to a point in $\mathbb{R}^n$ and two vertices $x \sim y$ are connected by a curve $e_{x,y}$. Such a drawing becomes more illustrative if the curves $e_{x,y}$ only intersect at their endpoints (in the vertices) and only there hit vertices: We call such drawings **crossing-free**. As an exercise, one proves (Exercise 1.1.6):

Proposition 1.1.24 *Every (finite) graph has a crossing-free drawing in $\mathbb{R}^n$ with $n \geq 3$.*

A **planar** drawing of a graph G is a crossing-free drawing of G in the plane (i.e., in $\mathbb{R}^2$). If such a drawing exists, the graph G is also called **planar**. See Fig. 1.12.

Planarity implies many other properties: This will be discussed in Chap. 4. For this, further graph operations[1] are used:

Definition 1.1.25

(a) An **edge subdivision** is an operation in which an edge is replaced by a path. I.e., an edge $e = \{x, y\}$ is replaced by edges $\{x_{j-1}, x_j\}$ with $j \in \{1, \ldots, m\}$, where $x_0 = x$, $x_m = y$ and the vertices x_j with $j = 1, \ldots, m - 1$ are new vertices. A graph G' is a **subdivision** of a graph G, if G' is obtained by edge subdivisions in G (Fig. 1.13).

(b) Another important operation is the **edge contraction**. If $e \in E$, $e = \{x_1, x_2\}$, then $G \cdot e$ denotes the graph obtained by "contracting" the edge e: x_1 and x_2 are replaced by a new vertex x_*, which is connected to those vertices that were previously connected to x_1 or x_2 (Fig. 1.14):

Fig. 1.12 Drawings of graphs

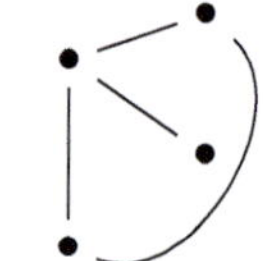

Edges do not intersect:
„good" (planar) drawing

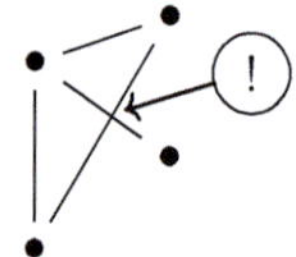

Edges intersect:
„bad" drawing

---

[1] Note for the lecturers: These terms are only used in Chap. 4 and can also be introduced there.

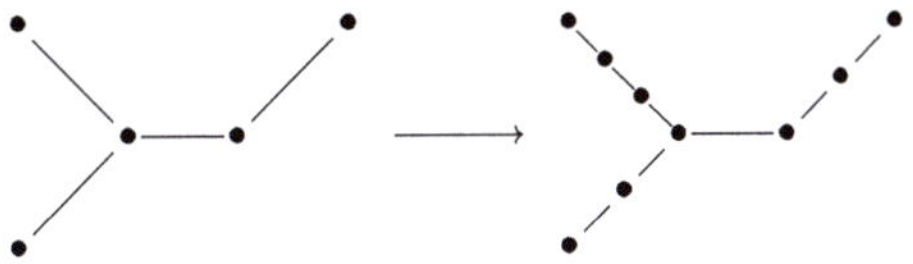

Fig. 1.13 Subdivision of a graph

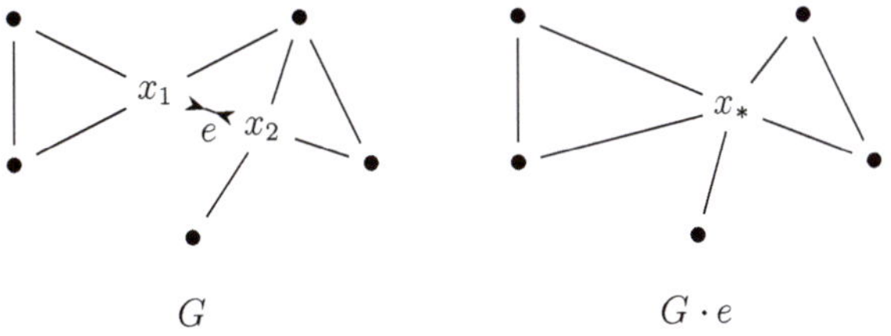

Fig. 1.14 Example of an edge contraction

$$G \cdot e := (X', E'), \qquad X' := \big(X \setminus \{x_1, x_2\}\big) \cup \{x_*\},$$

$$E' := E \setminus \{e \in E : e \sim x_1 \text{ or } e \sim x_2\}$$

$$\cup \big\{ \{x_*, y\} : \{y, x_1\} \in E \text{ or } \{y, x_2\} \in E \big\}.$$

A simple argument shows that $G \cdot e$ is connected if and only if G is connected.

(c) A graph G' is a **minor** of a graph G if G' can be obtained from G through successive applications of vertex removals, edge removals, and edge contractions.

Exercise 1.1.1 Let $G = (X, E)$ be a tree with $|X| \geq 2$. Show:

1. The graph G has at least two pendant vertices.
    Hint: Consider the longest path in G.
2. If you remove a pendant vertex from G, you get a tree again.

Exercise 1.1.2 Let $G = (X, E)$ be a connected graph. An edge $e \in E$ is called a *bridge* of G if the graph $G \setminus e$ is not connected.

1. Show: A connected graph is a tree if and only if each of its edges is a bridge.
2. Deduce that every connected graph has a spanning tree.

Exercise 1.1.3 Let $G = (X, E)$ be a graph. Show that the following statements are equivalent:

(a) G is a tree,
(b) G is connected and $|X| = |E| + 1$.
(c) G is cycle-free and $|X| = |E| + 1$.

Hint: Use Exercises 1.1.1 and 1.1.2.

Exercise 1.1.4 Let $G = (X, E)$ be a graph and let $x, y \in X$ with $x \neq y$. Show in detail: If there is a walk from x to y, then there is also a path from x to y.

Exercise 1.1.5 Show that every tree is a bipartite graph.

Hint: Let $G = (X, E)$ be a tree. Choose a fixed vertex $a \in X$ and consider the vertices x such that the distance between x and a is an even or odd number.

Exercise 1.1.6 Let $n \geq 3$. Show that every graph can be drawn in $\mathbb{R}^n$ without crossings.

Hint: One can place all vertices on a straight line.

1.2   Multiple Connectivity

The terms introduced above are essentially part of general mathematical culture and are also well known outside of graph theory. We will now discuss some more specific constructions.

The concept of a connected graph can be strengthened as follows. The associated terms will play a central role in the final chapters.

Definition 1.2.1 (Multiply Connected Graphs) Let $k \in \mathbb{N}$ and $G := (X, E)$ be a graph.

- A set $Y \subsetneq X$ is a **separating vertex set**, if $G \setminus Y$ is not connected <u>or</u> consists of a single vertex.

  The graph G is called **k-connected** (or also **k-vertex-connected**), if every separating vertex set has at least k elements.

  In other words: The graph G is k-connected, if it has at least $k + 1$ vertices and remains connected even after the removal of any $\leq (k - 1)$ vertices.
- A set $E' \subset E$ is a **separating edge set**, if $G \setminus E'$ is not connected.

  The graph G is called **k-edge-connected**, if $|X| \geq 2$ and every separating edge set has at least k elements.

  In other words: The graph G is k-edge-connected, if it has at least 2 vertices and remains connected even after the removal of any $k - 1$ edges.

We note that the conditions

- G has at least two vertices and is connected,
- G is 1-connected ($\equiv$ 1-vertex-connected),
- G is 1-edge-connected,

are equivalent by definition. The disconnected graphs as well as the graph consisting of a single vertex are often referred to as 0-connected.

The following non-trivial theorem provides an important property of k-connected graphs.[2]

Theorem 1.2.2 (Menger) *Two paths from a vertex u to a vertex v are called internally vertex-disjoint if they have no common vertices except u and v.*

If $G = (X, E)$ is a k-connected graph, then there are at least k paths between any vertices $u, v \in X$ with $u \neq v$ that are pairwise internally vertex-disjoint.

Proof We perform induction on k. For $k = 1$, the statement is true.

Now let $k \in \mathbb{N}$ such that in all k-connected graphs any two vertices are connected by at least k pairwise internally vertex-disjoint paths, and let $G = (X, E)$ be a $(k + 1)$-connected graph. Let $u, v \in X$. By an *A-family* we understand $k + 1$ paths $P_1, \ldots, P_k, P$, from u to v in G,

$$P_j = (u, p_j, \ldots, v), \quad j \in \{1, \ldots, k\}, \quad P = (u, p_{k+1}, \ldots, v),$$

such that

- the paths $P_1, \ldots, P_k$ are pairwise internally vertex-disjoint and
- the vertices $p_1, \ldots, p_{k+1}$ are all different.

By the *pivot vertex* of the A-family we understand the first vertex $\neq u$ on P that lies on one of the $P_1, \ldots, P_k$.

We will first show that A-families really exist. Since G is also k-connected, there are k pairwise internally vertex-disjoint paths $P_1, \ldots, P_k$ from u to v, $P_j = (u, p_j, \ldots, v)$, by the induction hypothesis. The graph $G \setminus \{p_1, \ldots, p_k\}$ is connected, so there is another path $P = (u, p_{k+1}, \ldots, v)$ in $G \setminus \{p_1, \ldots, p_k\}$ and thus also in G, such that $p_{k+1} \notin \{p_1, \ldots, p_k\}$. Thus, $(P_1, \ldots, P_k, P)$ is an A-family. Now we choose the A-family $(P_1, \ldots, P_k, P)$ such that the distance between the pivot vertex x of this family and v in $G \setminus u$ is smallest. Let P_{k+1} be the portion of P between u and x.

If $x = v$, we have constructed $k + 1$ internally vertex-disjoint paths $P_1, \ldots, P_{k+1}$ from u to v. From now on, let $x \neq v$.

The graph $G \setminus x$ is k-connected. By the induction hypothesis, we find in $G \setminus x$ pairwise internally vertex-disjoint paths $Q_1, \ldots, Q_k$ from u to v,

$$Q_j = (u, q_j, \ldots, v), \quad j \in \{1, \ldots, k\}.$$

We can additionally assume that the paths Q_j are chosen so that they contain the smallest possible number of edges in

---

[2] Note for the instructors: Menger's theorem is used in our book only in Exercise 1.2.3 and in Sect. 4.5.

$$E' := E \setminus \{\text{edges in } P_1, \ldots, P_{k+1}\}$$

Let now H be the graph consisting of the vertices and edges contained in $Q_1, \ldots, Q_k$ as well as the vertex x. Then there is a $j \in \{1, \ldots, k+1\}$ such that $p_j \notin \{q_1, \ldots, q_k\}$, i.e. the edge $\{u, p_j\}$ is not in H. Let y be the first vertex of P_j that lies in H. If $y = v$, we have constructed $k + 1$ internally vertex-disjoint paths $Q_1, \ldots, Q_k, P_j$ from u to v in G. From now on, let $y \neq v$.

If $y = x$, let R denote the shortest path from x to v in $G \setminus u$. Let z be the first vertex of R that lies on one of the Q_i. Then $(Q_1, \ldots, Q_k, P_j \cup R)$ is an A-family with pivot vertex z and the distance in $G \setminus u$ from z to v is strictly less than from x to v, which contradicts the above choice of $(P_1, \ldots, P_k, P)$. Thus $y \neq x$ and there exists an $i \in \{1, \ldots, k\}$ such that y lies in Q_i.

Now consider the section $(u = u_0, \ldots, u_m = y)$ of Q_i between u and y. Assume that each edge $\{u_{i-1}, u_i\}$ with $i \in \{1, \ldots, m\}$ lies in one of the $P_1, \ldots, P_{k+1}$. Two cases need to be examined:

(a) There exists an $\ell \in \{1, \ldots, k+1\}$ such that $\{u_{i-1}, u_i\}$ for all $i \in \{1, \ldots, m\}$ lies in P_ℓ. Thanks to the choice of y, $\ell \neq j$. Thus, P_ℓ and P_j intersect at y, which is impossible because $y \notin \{x, u, v\}$.

(b) There exist $\ell, \ell' \in \{1, \ldots, k+1\}$ with $\ell \neq \ell'$ and an $i \in \{1, \ldots, m-1\}$ such that $\{u_{i-1}, u_i\}$ lies in P_ℓ and $\{u_i, u_{i+1}\}$ lies in $P_{\ell'}$.
Then P_ℓ and $P_{\ell'}$ intersect at the vertex u_i. Since $Q_1, \ldots, Q_k$ are paths in $G \setminus x$, $u_i \neq x$. Furthermore, $u_i \notin \{u, v\}$ by construction. Since any two paths among $P_1, \ldots, P_{k+1}$ can only intersect in u, v, x, this case is also impossible.

Thus, there must exist an $i \in \{1, \ldots, m\}$ such that the edge $\{u_{i-1}, u_i\}$ lies in none of the $P_1, \ldots, P_{k+1}$, i.e., $\{u_{i-1}, u_i\} \in E'$. Construct a new path $\tilde{Q}_i$ from u to v by replacing the section of Q_i between u and y with the section of P_j between u and y. Note that this section of P_j does not contain the vertex x and does not intersect any of the paths $Q_1, \ldots, Q_k$. Thus, $\tilde{Q}_i$ is again a path from u to v in $G \setminus x$ and thus $Q_1, \ldots, Q_{i-1}, \tilde{Q}_i, Q_{i+1}, \ldots, Q_k$ are pairwise internally vertex-disjoint paths from u to v in $G \setminus x$. This new family, however, contains strictly fewer edges in E' than the original family $Q_1, \ldots, Q_k$. This contradicts the above choice of the $Q_1, \ldots, Q_k$. $\qquad\square$

As a corollary, we obtain:

Theorem 1.2.3 (Whitney)   *A k-connected graph is also k-edge-connected.*

Proof Let $G = (X, E)$ be a k-connected graph and let $E' \subset E$ with $|E'| \leq k - 1$. Let $u, v \in X$ with $u \neq v$. According to Menger's theorem (Theorem 1.2.2), there are at least k pairwise internally vertex-disjoint paths $P_1, \ldots, P_k$ from u to v in G. Each edge in E' lies in at most one of these paths. Since $|E'| \leq k - 1$, there exists an $i \in \{1, \ldots, k\}$ such that the path P_i does not contain any edge from E' and thus connects u and v in $G \setminus E'$. It follows that u and v lie in the same connected component of $G \setminus E'$. Since u and v were chosen arbitrarily, $G \setminus E'$ is connected. $\quad\square$

Whitney's theorem can also be proven directly, without using Menger's theorem: see Exercise 1.2.1.

Remark 1.2.4 There are also versions of Menger's theorem for graphs with multiple edges, see e.g. the book by Diestel [45, Sec. 2.3]. The above formulation and proof follow the article by McCuaig [80].

Exercise 1.2.1 We want to prove Whitney's theorem (Theorem 1.2.3) directly (independently of Menger's theorem).

Let $G = (X, E)$ be a k-vertex-connected graph and let E' be a separating edge set such that $|E'|$ is minimal. Denote $n := |X|$.

1. Show: There exists a subset $S \subset X$ (with $S \neq \varnothing$ and $S \neq X$) such that

$$E' = \{e \in E : e \text{ connects a vertex in } S \text{ with a vertex in } X \setminus S\}.$$

   Hint: Consider a connected component of $G \setminus E'$.
   Now we consider two complementary cases:
   (F1) For all $x \in S$ and $y \in X \setminus S$ we have $\{x, y\} \in E$.
   (F2) There exist $x \in S$ and $y \in X \setminus S$ with $\{x, y\} \notin E$.
2. Assume (F1) is satisfied. Show that $|E'| \geq n - 1$, and deduce that G is k-edge-connected in this case.
   From now on, assume (F2) is satisfied. We consider

$$X' := \left\{ y' \in X \setminus S : y' \sim x \right\} \cup \left\{ x' \in S \setminus \{x\} : \exists z \in X \setminus S \text{ with } x' \sim z \right\}.$$

3. Show that $x, y \in X \setminus X'$ and that $G \setminus X'$ contains no path from x to y.
4. Show that $|X'| \leq |E'|$.
   Hint: Construct an injective mapping $X' \to E'$.
5. Deduce that G is a k-edge-connected graph.

Exercise 1.2.2 Show in detail that trees are not 2-connected.

Exercise 1.2.3 (Used in Sect. 4.5) Let $G = (X, E)$ be a k-connected graph (with $k \geq 1$) and let $x_1, \ldots, x_k \in X$ be distinct vertices.

1. We construct a new graph $\widetilde{G}$ by adding a new vertex that is adjacent to $x_1, \ldots, x_k$. Show that $\widetilde{G}$ is also a k-connected graph.
2. Let $y \in X \setminus \{x_1, \ldots, x_k\}$. Show that there are paths P_i from y to x_i, with $i \in \{1, \ldots, k\}$, such that for any $i \neq j$ the only common vertex of P_i and P_j is y.

1.3 Important Graph Invariants

In this section, we will introduce some of the most well-known graph invariants. These definitions will be important for the rest of the book.

Let $G = (X, E)$ be a graph. Continuing the discussion of multiple connectivity, define:

- the **vertex connectivity** $\kappa(G)$ of G,

$$\kappa(G) := \min\left\{|Y| :\ Y \subsetneq X \text{ is a separating vertex set}\right\}$$

$$\equiv \max\{k \in \mathbb{N} \cup \{0\} :\ G \text{ is } k\text{-connected}\},$$

- the **edge connectivity** $\kappa'(G)$ of G,

$$\kappa'(G) := \min\left\{|E'| :\ E' \subset E \text{ is a separating edge set}\right\}$$

$$= \max\{k \in \mathbb{N} \cup \{0\} :\ G \text{ is } k\text{-edge-connected}\}.$$

Whitney's theorem (Theorem 1.2.3) can equivalently be reformulated as

$$\kappa(G) \leq \kappa'(G).$$

Further inequalities are discussed in Sect. 2.7.

Let C be a non-empty set. A **vertex coloring** on G is a mapping $c : X \to C$, such that

$$c(x) \neq c(y) \text{ for all } x, y \in X \text{ with } x \sim y.$$

The elements of C are often interpreted as colors: Each vertex $x \in X$ is assigned a color $c(x)$ and it is required that adjacent vertices always have different colors (Fig. 1.15).

Definition 1.3.1 The **chromatic number** $\chi(G)$ of a graph $G = (X, E)$ is

$$\chi(G) := \min\left\{|C| :\ \text{there exists a vertex coloring } c : X \to C\right\}.$$

Various estimates for the chromatic number are considered in Sect. 2.6.

Fig. 1.15 Vertex colorings

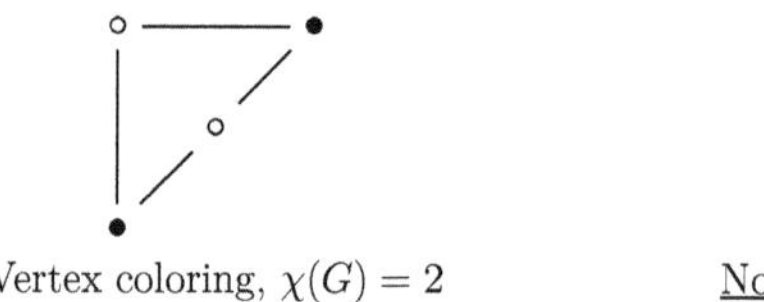

Vertex coloring, $\chi(G) = 2$

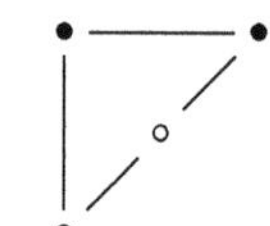

Not a vertex coloring

Remark 1.3.2 The chromatic number is closely related to bipartite graphs. If G is bipartite (with partition components Y and Y'), there are two possible cases:

- G has no edges ($E = \varnothing$). In this case, it is clear that $\chi(G) = 1$.
- G has at least one edge. There is no vertex coloring with only one color, as the end vertices of each edge must have different colors. However, there is a vertex coloring with two colors (one color for Y and a different color for Y'), so $\chi(G) = 2$.

Conversely:

- Every graph $G = (X, E)$ with $\chi(G) = 1$ has no edges ($E = \varnothing$) and is therefore bipartite.
- If $G = (X, E)$ is a graph with $\chi(G) = 2$, the subsets

$$Y := \text{the vertices of the first color,}$$

$$Y' := \text{the vertices of the second color}$$

form a decomposition of X into two parts with the required properties, so in this case G is also bipartite.

Overall, we have the following equivalences:

- $\chi(G) \leq 2$ if and only if G is bipartite.
- $\chi(G) = 2$ if and only if G is bipartite <u>and</u> has at least one edge.

Additional characterizations of bipartite graphs can be found in Sect. 2.5.

Another group of graph invariants is related to special subgraphs. A subset of vertices $Y \subset X$ is called

- a **clique** in G, if the induced subgraph G_Y is complete (i.e., all vertices of Y are adjacent to each other),
- an **anticlique** in G (or an **independent set of vertices**), if the induced subgraph G_Y is empty (i.e., there is no edge between the vertices of Y).

Searching for cliques is an important problem in the study of social networks (a clique can be interpreted as a "community" around a topic). The number of vertices in an anticlique provides a lower estimate for the number of communities, as the vertices of an anticlique obviously belong to different cliques. We define:

- the **clique number** $\omega(G)$ of G,

$$\omega(G) = \text{the number of vertices in the largest clique in } G$$

$$\equiv \max\left\{ |Y| : Y \subset X \text{ clique in } G \right\},$$

- the **anticlique number** (or the **independence number**) $\alpha(G)$ of G,

$$\alpha(G) = \text{the number of vertices in the largest anticlique in } G$$

$$\equiv \max\{|Y| : Y \subset X \text{ anticlique in } G\}.$$

For every vertex $x \in X$, the subgraph $(\{x\}, \varnothing) \subset G$ is a clique and at the same time an anticlique, so $\omega(G) \geq 1$ and $\alpha(G) \geq 1$ for all G. Directly from the definition follows:

Proposition 1.3.3 *Let $G = (X, E)$ be a graph. A set $Y \subset X$ is a clique in G if and only if it is an anticlique in the complement graph $\overline{G}$. In particular,*

$$\omega(G) = \alpha(\overline{G}), \quad \alpha(G) = \omega(\overline{G}).$$

The two new invariants are related to $\chi(G)$.

Proposition 1.3.4 *For every graph $G = (X, E)$, the inequalities*

$$\chi(G) \geq \omega(G), \qquad \alpha(G)\,\chi(G) \geq |X|.$$

hold.

Proof Let Y be a clique with $|Y| = \omega(G)$. In any vertex coloring of G, all vertices in Y must have different colors, so $\chi(G) \geq |Y| = \omega(G)$.

Let $k = \chi(G)$ and $c : X \to \{1, \ldots, k\}$ be a vertex coloring, then each of the pairwise disjoint sets

$$X_j := \{x \in X : c(x) = j\} \subset X, \quad j \in \{1, \ldots, k\},$$

is an anticlique, so $|X_j| \leq \alpha(G)$ for each j, and it follows

$$|X| = \left|\bigcup_{j=1}^{k} X_j\right| = \sum_{j=1}^{k} |X_j| \leq k\,\alpha(G) = \chi(G)\,\alpha(G).$$

$\square$

In Sects. 2.8 and 2.9, we show several more inequalities for the clique and anticlique numbers.

Later in the text, we will also introduce and investigate other graph invariants, in particular:

- the number $N(G)$ of spanning trees in G (Sect. 1.8),

- the Cheeger constants $h(G)$ and $h'(G)$, which take non-negative real values and can be used to characterize the "minimum density" of G (Sect. 3.1),
- the Colin de Verdière invariant $\mu(G)$ with values in $\mathbb{N}_0 := \mathbb{N} \cup \{0\}$: This can be used, in particular, to characterize planar graphs (Sect. 4.4).

We will soon see how the above-mentioned and some other graph invariants can be estimated using the eigenvalues and eigenfunctions of graph matrices.

1.4   Linear Mappings, Matrices, Eigenvalues

We will now briefly review selected topics from linear algebra. This review is not intended as a replacement for a basic lecture in linear algebra: it is only meant to provide orientation on which topics of linear algebra are actively used in the book (and which should be refreshed if necessary).

Remark 1.4.1 Although in spectral graph theory one usually works with real matrices, sometimes intermediate calculations in the complex domain are required (e.g. in Sects. 1.6 and 3.5). Therefore, we denote

$$\mathbb{K} := \mathbb{R} \text{ or } \mathbb{C}$$

and make a specific choice depending on the context. In this book, we always understand a vector space to be a <u>finite</u>-dimensional $\mathbb{K}$-vector space (unless otherwise explicitly assumed). Similarly, a linear mapping is understood to be a $\mathbb{K}$-linear mapping.

Definition 1.4.2 Let V be a vector space and let $M : V \to V$ be a linear mapping. An **eigenvector** (or an **eigenfunction**) of M is a vector $v \in V$ with $v \neq 0$, such that $Mv = \lambda v$ for some $\lambda \in \mathbb{K}$. The number λ is called the **eigenvalue** of M for v. The subspace $\ker(M - \lambda I) \subset V$ is called the **eigenspace** of M for λ, and its dimension is called the **multiplicity** of λ. If the multiplicity is 1, the eigenvalue λ is called simple.

Remark 1.4.3 Often one considers the case $V := \mathbb{K}^n$ with $n \in \mathbb{N}$. A linear mapping $M : \mathbb{K}^n \to \mathbb{K}^n$ is then identified with the $n \times n$ matrix $M = (M_{j,k})$, if one views Mx as column multiplication,

$$(Mx)_j := \sum_{k=1}^{n} M_{j,k} x_k, \quad j \in \{1, \ldots, n\}.$$

From linear algebra it is known that a number $\lambda \in \mathbb{K}$ is an eigenvalue of M if and only if $\det(\lambda I - M) = 0$, where I is the identity matrix. The function

$$P_M : \; \mathbb{K} \ni t \mapsto \det(tI - M)$$

is a polynomial, which is referred to as the **characteristic polynomial** of M. It holds

$$P_M(t) = \sum_{k=0}^{n} B_k t^{n-k},$$

where $B_0 := 1$ and

$$B_s = (-1)^s \cdot \text{the sum of the principal minors of } M \text{ of order } s \text{ for } s \geq 1.$$

In particular

$$B_1 = -\sum_{j=1}^{n} M_{j,j} = -\operatorname{tr} M, \quad B_n = (-1)^n \det M.$$

If $\lambda_1, \ldots, \lambda_n$ are the complex roots of P_M, then

$$P_M(t) = (t - \lambda_1) \cdot \ldots \cdot (t - \lambda_n) = t^n + \sum_{k=1}^{n} (-1)^k \Big(\sum_{1 \leq n_1 < \cdots < n_k \leq n} \lambda_{n_1} \cdot \ldots \cdot \lambda_{n_k} \Big) t^{n-k}.$$

By comparing coefficients, we obtain for $k \geq 1$:

$$(-1)^k \sum_{1 \leq n_1 < \cdots < n_k \leq n} \lambda_{n_1} \cdot \ldots \cdot \lambda_{n_k} = B_k.$$

In particular, the identities

$$\lambda_1 + \cdots + \lambda_n = \operatorname{tr} M, \quad \lambda_1 \cdot \ldots \cdot \lambda_n = \det M,$$

often referred to as **trace theorems**, hold.

Remark 1.4.4 If in the above remark the matrix $(M_{j,k})$ is diagonal, i.e. $M_{j,k} = 0$ for all $j \neq k$, then we have the characteristic polynomial

$$p_M(t) = (t - M_{1,1}) \cdot \ldots \cdot (t - M_{n,n}).$$

The roots of p_M (and the eigenvalues of M) are exactly the diagonal coefficients $M_{j,j}$ of M.

Remark 1.4.5 Let V be a vector space of dimension $n \in \mathbb{N}$ and $\mathcal{V} := (v_1, \ldots, v_n)$ a basis in V. Let $M : V \to V$ be a linear mapping. The representation matrix $M_\mathcal{V}$ of M in the basis $\mathcal{V}$ is

$$M_{\mathcal{V}} := (M_{j,k})_{j,k=1}^{n},$$

which is uniquely defined by the following property: If $v \in V$ with

$$v = \sum_{j=1}^{n} x_j v_j, \quad x_j \in \mathbb{K}, \tag{1.1}$$

then

$$Mv = \sum_{j=1}^{n} y_j v_j \text{ with } y_j = \sum_{k=1}^{n} M_{j,k} x_k,$$

in other words,

$$\begin{pmatrix} y_1 \\ \dots \\ y_n \end{pmatrix} = M_{\mathcal{V}} \begin{pmatrix} x_1 \\ \dots \\ x_n \end{pmatrix}.$$

The characteristic polynomial p_M is defined as

$$p_M(t) := p_{M_{\mathcal{V}}}(t) \equiv \det(tI - M_{\mathcal{V}}).$$

The matrix $M_{\mathcal{V}}$ depends on the choice of the basis $\mathcal{V}$, but not the characteristic polynomial: If $\mathcal{U}$ is another basis of V, then $M_{\mathcal{U}} = S^{-1} M_{\mathcal{V}} S$ with an invertible matrix S (transition matrix), hence

$$\det(tI - M_{\mathcal{U}}) = \det\left(tI - S^{-1} M_{\mathcal{V}} S\right) = \det\left(S^{-1}(tI - M_{\mathcal{V}})S\right)$$

$$= \det(S^{-1}) \det(tI - M_{\mathcal{V}}) \det S = \frac{1}{\det S} \det(tI - M_{\mathcal{V}}) \det S$$

$$= \det(tI - M_{\mathcal{V}}).$$

A vector $v \in V$ is an eigenvector of M with eigenvalue λ if and only if the column $\begin{pmatrix} x_1 \\ \dots \\ x_n \end{pmatrix}$ with x_j in (1.1) is an eigenvector of the matrix $M_{\mathcal{V}}$ with the same eigenvalue. It follows that the eigenvalues of M are exactly the roots of the characteristic polynomial p_M in $\mathbb{K}$. Since every polynomial in $\mathbb{K}$ has at least one complex root according to the Fundamental Theorem of Algebra, every linear mapping has at least one eigenvalue in $\mathbb{C}$.

If $\mathcal{V}$ is an **eigenbasis** for M, i.e., if each v_j is an eigenvector of M, with $Mv_j = \lambda_j v_j$, then the representation matrix $M_{\mathcal{V}}$ is the diagonal matrix with λ_j on the diagonal. It follows (Remark 1.4.4) that the set of eigenvalues of M coincides with the set $\{\lambda_j : j \in \{1, \dots, n\}\}$.

In the same way one proves:

Proposition 1.4.6 *Let V, V' be vector spaces and $M : V \to V$ and $M' : V' \to V'$ be similar linear mappings, i.e. $SM = M'S$ for a bijective linear mapping $S : V \to V'$. Then*

$$\dim \ker(M - \lambda I) = \dim \ker(M' - \lambda I) \text{ for all } \lambda \in \mathbb{K}.$$

In particular, M and M' have the same eigenvalues with the same multiplicities.

Now let V be a vector space with a scalar product $\langle \cdot, \cdot \rangle$. If $\mathbb{K} = \mathbb{C}$, it is always assumed that the scalar product $\langle u, v \rangle$ is conjugate linear with respect to u and linear with respect to v. For example, the usual scalar product in $\mathbb{K}^n$ is then given by

$$\langle u, v \rangle := \sum_{j=1}^{n} \overline{u_j}\, v_j, \quad u = (u_1, \ldots, u_n), \quad v = (v_1, \ldots, v_n),$$

As a reminder: This gives rise to the induced norm $\| \cdot \|$ on V,

$$\|v\| := \sqrt{\langle v, v \rangle}, \quad v \in V,$$

and we have the Cauchy-Schwarz inequality

$$\left| \langle u, v \rangle \right| \leq \|u\| \, \|v\|, \quad u, v \in V,$$

where equality holds if and only if u and v are linearly dependent. A vector $v \in V$ is called **normalized**, if $\|v\| = 1$.

For $u, v \in V$ we write $u \perp v$ if $\langle u, v \rangle = 0$: In this case, u and v are called **orthogonal**. If $U, U' \subset V$ are subspaces, we write $U \perp U'$ if $u \perp u'$ for all $u \in U$ and all $u' \in U'$. For a set $U \subset V$, the **orthogonal complement** $U^\perp$ of U in V is defined by

$$U^\perp := \left\{ v \in V : v \perp u \text{ for all } u \in U \right\}.$$

For $u \in V$ we denote

$$u^\perp := \{u\}^\perp.$$

An **orthonormal system** in V is a family of vectors $\mathcal{V} := (v_1, \ldots, v_k)$ such that $\|v_j\| = 1$ for all j and $v_i \perp v_j$ for all $i \neq j$. If this is also a basis, then the family is called an **orthonormal basis**: This is exactly the case when $\mathcal{V}^\perp = \{0\}$. If $\mathcal{V}$ is an orthonormal basis, then for each $v \in V$ we have the representation

$$v = \sum_{j=1}^{n} x_j v_j, \quad x_j := \langle v_j, v \rangle, \qquad \|v\|^2 = \sum_{j=1}^{n} |x_j|^2.$$

If $M : V \to V$ is a linear mapping, then its representation matrix $M_{\mathcal{V}}$ in the orthonormal basis $\mathcal{V}$ has the explicit form $M_{\mathcal{V}} := \big(\langle v_i, M v_j \rangle\big)$. From linear algebra, it is known that every finite-dimensional vector space has a basis and this can be transformed into an orthonormal basis using the Gram-Schmidt orthogonalization process, if there is a scalar product.

Now let U, V be two vector spaces with scalar products and let $M : U \to V$ be a linear mapping. The **adjoint** linear mapping $M^* : V \to U$ is defined by

$$\langle Mu, v \rangle = \langle u, M^* v \rangle \text{ for all } u \in U,\ v \in V.$$

If W is another vector space and $N : V \to W$ is a linear mapping, then it follows from the definition that $(NM)^* = M^* N^*$ and $(M^*)^* = M$. If $U = V$ and $M = M^*$, M is called **self-adjoint** in U.

We will list some important cases of adjoint and self-adjoint linear mappings explicitly.

Remark 1.4.7 (Multiplication by Matrices) Let M be an $m \times n$ matrix (with coefficients in $\mathbb{K}$). Consider the linear mapping

$$\mathbb{K}^n \ni u \mapsto Mu \in \mathbb{K}^m.$$

A simple calculation shows that the adjoint linear mapping is given by

$$\mathbb{K}^m \ni v \mapsto M^* v \in \mathbb{K}^n.$$

where M^* is the Hermitian transposed matrix to M, i.e. $(M^*)_{j,k} = \overline{M_{k,j}}$ for $(j, k) \in \{1, \ldots, n\} \times \{1, \ldots, m\}$. Thus, the original mapping is self-adjoint exactly when the matrix M is Hermitian, i.e. $m = n$ and $M_{j,k} = \overline{M_{k,j}}$ for all j, k. If M is a real matrix, M^* coincides with the transposed matrix $M^\top$, but we prefer to use the uniform complex notation M^* also for real matrices M.

In particular, every real symmetric $n \times n$ matrix defines a self-adjoint linear mapping in $\mathbb{C}^n$ and $\mathbb{R}^n$.

Remark 1.4.8 (Orthogonal Projections and Embeddings) Let V be a subspace in U and $P : U \to V$ the orthogonal projection, i.e., for each $u \in U$, Pu is the only vector in V with $u - Pu \perp V$. The adjoint operator P^* is the natural embedding of V into U,

$$P^* : V \ni v \mapsto v \in U,$$

and $PP^* : V \to V$ is the identity mapping.

We will now prove the most important properties of self-adjoint linear mappings (our experience has shown that these constructions are rather poorly covered in basic lectures and are hardly known to most students).

Lemma 1.4.9 *Let V be a vector space with scalar product and let $M : V \to V$ be a self-adjoint linear mapping, then:*

(a) All eigenvalues of M are real.
(b) If λ and μ are two distinct eigenvalues of M, then

$$\ker(M - \lambda I) \perp \ker(M - \mu I).$$

Proof

(a) For the case $\mathbb{K} = \mathbb{R}$ the statement is trivial (since all eigenvalues of a $\mathbb{K}$-linear mapping are by definition in $\mathbb{K}$). From now on let $\mathbb{K} = \mathbb{C}$ and let $\lambda \in \mathbb{C}$ be an eigenvalue of M, then $Mv = \lambda v$ for a $v \in V$ with $v \neq 0$. Then we have

$$\lambda \langle v, v \rangle = \langle v, \lambda v \rangle = \langle v, Mv \rangle = \langle Mv, v \rangle = \langle \lambda v, v \rangle = \overline{\lambda} \langle v, v \rangle.$$

Since $v \neq 0$ we have $\langle v, v \rangle \neq 0$, thus $\lambda = \overline{\lambda}$, i.e. $\lambda \in \mathbb{R}$.
(b) From (a) it follows that λ and μ are real. Let $u \in \ker(M - \lambda I)$ and $v \in \ker(M - \mu I)$, i.e. $Mu = \lambda u$ and $Mv = \mu v$. We have

$$\lambda \langle u, v \rangle = \langle \lambda u, v \rangle = \langle Mu, v \rangle = \langle u, Mv \rangle = \langle u, \mu v \rangle = \mu \langle u, v \rangle,$$

i.e. $(\lambda - \mu)\langle u, v \rangle = 0$, and from $\lambda \neq \mu$ it follows $\langle u, v \rangle = 0$.

$\square$

Definition 1.4.10 Let V be a vector space and $M : V \to V$ a linear mapping. An **invariant subspace** of M is a subspace $U \subset V$ with $M(U) \subset U$.

Lemma 1.4.11 *Let V be a vector space with scalar product, $M : V \to V$ a self-adjoint linear mapping, $U \subset V$ an invariant subspace of M. Then $U^{\perp}$ is also an invariant subspace of M. In particular, it follows that M also defines a self-adjoint linear mapping in $U^{\perp}$.*

Proof Let $v \in U^{\perp}$, then for all $u \in U$ we have $\langle u, Mv \rangle = \langle \underbrace{Mu}_{\in U}, v \rangle = 0$, i.e. $Mv \in U^{\perp}$.

$\square$

From these considerations we deduce one of the most important results about self-adjoint linear mappings, which plays a central role throughout the book:

Theorem 1.4.12 (Spectral Theorem) *Let V be a (finite-dimensional) vector space with scalar product and let $M : V \to V$ be a self-adjoint linear mapping. Then:*

(a) All eigenvalues of M are real.

(b) There is an orthonormal basis of V consisting of eigenvectors of M.
    More precisely: let $n := \dim V$, then there are $v_1, \ldots, v_n \in V$ and $\lambda_1, \ldots, \lambda_n \in \mathbb{R}$, such that

$$\langle v_j, v_k \rangle = \delta_{j,k}, \quad M v_j = \lambda_j v_j \quad \text{for all } j, k \in \{1, \ldots, n\}.$$

Proof Let $\mu_1, \ldots, \mu_N$ be the pairwise distinct eigenvalues of M: They are real according to Lemma 1.4.9. Consider the associated eigenspaces $U_j := \ker(M - \mu_j I)$, $j \in \{1, \ldots, N\}$, which are pairwise orthogonal according to Lemma 1.4.9, and denote

$$U := U_1 \oplus \cdots \oplus U_N.$$

If $u \in U$, then $u = u_1 + \cdots + u_N$ with $u_j \in U_j$. For each j, $M u_j = \mu_j u_j \in U_j$, so

$$Mu = M u_1 + \cdots + M u_N = \mu_1 u_1 + \cdots + \mu_N u_N \in U,$$

i.e. $M(U) \subset U$. According to Lemma 1.4.11, $M : U^\perp \to U^\perp$ is a self-adjoint linear mapping in $U^\perp$. If $U^\perp \neq \{0\}$, M must have at least one eigenvector $v \in U^\perp$ with eigenvalue μ. Then μ must be among the $\mu_1, \ldots, \mu_N$ and v lies in one of the U_j and thus in U. So $v \in U^\perp \cap U = \{0\}$ and $v = 0$: Contradiction! Therefore $U^\perp = \{0\}$, from which follows $U = V$, i.e.

$$V = U_1 \oplus \cdots \oplus U_N. \tag{1.2}$$

Denote $n_j := \dim U_j$ and let $u_{j,1}, \ldots, u_{j,n_j}$ be an orthonormal basis in U_j. According to (1.2),

$$\mathcal{E} := (u_{j,k})_{j \in \{1,\ldots,N\}, \, k \in \{1,\ldots,n_j\}}$$

is a generating system for V. Since the U_j are orthogonal to each other, $\mathcal{E}$ is an orthonormal system, i.e., $\mathcal{E}$ is an orthonormal basis in V, and $M u_{j,k} = \mu_j u_{j,k}$ by construction. By renumbering

$$v_1 := u_{1,1}, \qquad \ldots, \qquad v_{n_1} := u_{1,n_1},$$

$$v_{n_1+1} := u_{2,1}, \qquad \ldots, \qquad v_{n_1+n_2} := u_{2,n_2},$$

$$v_{n_1+\cdots+n_{N-1}+1} := u_{N,1}, \qquad \ldots, \qquad v_n \equiv v_{n_1+\cdots+n_N} := u_{N,n_N},$$

and

$$\lambda_1 = \ldots = \lambda_{n_1} := \mu_1,$$

$$\lambda_{n_1+1} = \ldots = \lambda_{n_1+n_2} := \mu_2,$$

$$\ldots$$

$$\lambda_{n_1+\cdots+n_{N-1}+1} = \ldots = \lambda_{n_1+n_2+\cdots+n_N} := \mu_N,$$

we obtain the required representation. $\qquad\square$

Definition 1.4.13 Let M be a self-adjoint linear mapping in a vector space V. Let $\mu_1, \ldots, \mu_N$ be the pairwise distinct eigenvalues of M and $n_j := \dim \ker(M - \mu_j I)$ the associated multiplicities. The **spectrum** of M is the multiset

$$\sigma(M) := \big\{ \underbrace{\mu_1, \ldots, \mu_1}_{n_1\text{-times}}, \underbrace{\mu_2, \ldots, \mu_2}_{n_2\text{-times}}, \ldots, \underbrace{\mu_N, \ldots, \mu_N}_{n_N\text{-times}} \big\}.$$

Remark 1.4.14 A real symmetric $n \times n$ matrix M defines self-adjoint mappings through $M_{\mathbb{R}} : \mathbb{R}^n \to \mathbb{R}^n$ as well as through $M_{\mathbb{C}} : \mathbb{C}^n \to \mathbb{C}^n$. Since the two mappings $M_{\mathbb{R}}$ and $M_{\mathbb{C}}$ have the same characteristic polynomial and all eigenvalues (=the roots of the characteristic polynomial) are real, $M_{\mathbb{R}}$ and $M_{\mathbb{C}}$ have the same sets of pairwise distinct eigenvalues $\{\mu_1, \ldots, \mu_N\}$. Furthermore, from the Gaussian elimination method, it follows that for each $j \in \{1, \ldots, N\}$,

$$\dim_{\mathbb{R}}(M_{\mathbb{R}} - \mu_j I) = \dim_{\mathbb{C}}(M_{\mathbb{C}} - \mu_j I).$$

In particular, $\sigma(M_{\mathbb{R}}) = \sigma(M_{\mathbb{C}})$, and an orthonormal basis of eigenvectors of $M_{\mathbb{R}}$ in $\mathbb{R}^n$ can be seen as an orthonormal basis of eigenvectors of $M_{\mathbb{C}}$ in $\mathbb{C}^n$ using the usual embedding of $\mathbb{R}$ in $\mathbb{C}$. Depending on the situation, it may be convenient to work with real or complex spaces, as the spectrum is the same in both cases.

Remark 1.4.15 Let $M : V \to V$ be a self-adjoint linear mapping and $v_1, \ldots, v_n$ an orthonormal basis of eigenvectors of M, $Mv_j = \lambda_j v_j$ with $\lambda_j \in \mathbb{R}$. Let q be a polynomial with real coefficients,

$$q : t \mapsto \sum_{k=0}^{m} q_k t^k, \quad q_k \in \mathbb{R},$$

then the linear mapping $q(M) : V \to V$ is also self-adjoint. For each $j \in \{1, \ldots, n\}$,

$$q(M)v_j = \sum_{k=0}^{m} q_k M^k v_j = \sum_{k=0}^{m} q_k \lambda_j^k v_j = q(\lambda_j)v_j,$$

so $v_1, \ldots, v_n$ is also an orthonormal basis of eigenvectors of $q(M)$ and

$$\sigma\big(q(M)\big) = \{q(\lambda_j) : j \in \{1, \ldots, n\}\} \equiv q\big(\sigma(M)\big).$$

Exercise 1.4.4 (Spectral Theorem for Unitary Mappings) Let V be a complex (finite-dimensional) vector space. A linear mapping $\Theta : V \to V$ is called *unitary* if $\Theta^*\Theta = \Theta\Theta^* = I$ (or equivalently: $\Theta : V \to V$ is bijective with $\Theta^{-1} = \Theta^*$).

1. Show: For every eigenvalue λ of Θ it holds $|\lambda| = 1$.
2. Let λ and μ be two distinct eigenvalues of Θ. Show that the corresponding eigenspaces are orthogonal.
3. Let U be an invariant subspace of Θ. Show:
   (a) $\Theta(U) = U$,
   (b) U is an invariant subspace of Θ^{-1},
   (c) $U^\perp$ is an invariant subspace of Θ.
4. Deduce that there is an orthonormal basis of V consisting of eigenvectors of Θ.

1.5 Applications to Graph Matrices

For a non-empty finite set X, we consider the $\mathbb{K}$-vector space

$$\mathbb{K}^X := \{f : X \to \mathbb{K}\}$$

with the scalar product

$$\langle f, g \rangle \equiv \langle f, g \rangle_{\mathbb{K}^X} := \sum_{x \in X} \overline{f(x)}\, g(x).$$

For $\mathbb{K} = \mathbb{R}$, the complex conjugation is not necessary (and has no effect), but we will keep the uniform complex notation in this case as well. The space $\mathbb{K}^X$ is obviously isomorphic to $\mathbb{K}^{|X|}$.

Let $G = (X, E)$ be a graph. The matrices A, D, L, Q for G defined in Sect. 1.1 can also be considered as linear mappings in $\mathbb{K}^X$:

$$A : \mathbb{K}^X \to \mathbb{K}^X, \qquad (Af)(x) := \sum_{y \in X} A_{x,y} f(y) \equiv \sum_{y:\, y \sim x} f(y),$$

and analogously

$$(Lf)(x) = (\deg x) f(x) - \sum_{y:\, y \sim x} f(y) \equiv \sum_{y:\, y \sim x} \big(f(x) - f(y)\big), \qquad (1.3)$$

$$(Qf)(x) = (\deg x) f(x) + \sum_{y:\, y \sim x} f(y) \equiv \sum_{y:\, y \sim x} \big(f(x) + f(y)\big). \qquad (1.4)$$

It is easy to check that the mappings A, D, L, Q are self-adjoint (since they are given by real symmetric matrices). It follows that the spectral theorem applies to all these matrices.

Remark 1.5.1 In analysis and mathematical physics, one often works with the Laplace operator Δ. If $f : \mathbb{R}^n \to \mathbb{C}$ is a sufficiently smooth function, Δf is defined by

$$\Delta f(x) = \sum_{k=1}^{n} \frac{\partial^2 f}{\partial x_k^2}(x), \quad x \in \mathbb{R}^n.$$

Using the Taylor expansion for f,

$$f(y) - f(x) = \sum_{k=1}^{n} \frac{\partial f}{\partial x_k}(x)(y_k - x_k) + \frac{1}{2} \sum_{j,k=1}^{n} \frac{\partial^2 f}{\partial x_j \partial x_k}(x)(y_j - x_j)(y_k - x_k) + \ldots$$

one obtains

$$\lim_{\varepsilon \to 0^+} \frac{2n}{\omega_n \varepsilon^{n+1}} \int_{|x-y|=\varepsilon} \big(f(x) - f(y)\big)\, \mathrm{d}s(y) = -\Delta f(x), \tag{1.5}$$

where $\mathrm{d}s$ and ω_n denote the hypersurface measure in $\mathbb{R}^n$ and the hypersurface volume of the unit sphere in $\mathbb{R}^n$, respectively (in particular $\omega_2 = 2\pi$ and $\omega_3 = 4\pi$). The representation (1.3) for L can be considered as a discrete version of (1.5). Later in the text, we will see further various analogies between graph matrices and differential operators.

Remark 1.5.2 It is easy to see that the spectra of the above-introduced graph matrices are invariant with respect to graph isomorphisms. Namely, let $G' = (X', E')$ be a graph isomorphic to G with isomorphism $\varphi : X \to X'$. Let $R \in \{A, L, Q\}$ and let R' be the corresponding matrix for G', then

$$R'_{\varphi(x),\varphi(y)} = R_{x,y} \text{ for all } x, y \in X. \tag{1.6}$$

Consider the linear mapping

$$\Theta : \mathbb{K}^{X'} \to \mathbb{K}^X, \quad (\Theta f)(x) := f\big(\varphi(x)\big) \text{ for all } x \in X \text{ and } f \in \mathbb{K}^{X'},$$

which is obviously bijective ($\Theta f \equiv 0$ is equivalent to $f \equiv 0$). For all $f \in \mathbb{K}^{X'}$ and $x \in X$, we have

$$(\Theta R' f)(x) = (R' f)\big(\varphi(x)\big) = \sum_{y' \in X'} R'_{\varphi(x),y'} f(y')$$

$$(\text{substitute } \varphi(y) \text{ instead of } y') = \sum_{y \in X} R'_{\varphi(x),\varphi(y)} f\big(\varphi(y)\big) \overset{(1.6)}{=} \sum_{y \in X} R_{x,y} f\big(\varphi(y)\big)$$

$$= \sum_{y \in X} R_{x,y}(\Theta f)(y) = (R\Theta f)(x).$$

In other words, $\Theta R' = R\Theta$ and $\sigma(R') = \sigma(R)$ according to Proposition 1.4.6. It also follows that $\sigma(R)$ is independent of the chosen order of the elements in X.

Remark 1.5.3 If $G_i = (X_i, E_i)$ with $i \in \{1, \ldots, N\}$ are the connected components of G, $G = G_1 \sqcup \cdots \sqcup G_N$, and A_i are the adjacency matrices of G_i, then the decompositions

$$\mathbb{K}^X = \mathbb{K}^{X_1} \oplus \cdots \oplus \mathbb{K}^{X_N}, \quad A = A_1 \oplus \cdots \oplus A_N,$$

hold, in particular $\sigma(A) = \sigma(A_1) \cup \cdots \cup \sigma(A_N)$. Analogous representations also apply to the matrices D, L and Q for G and G_i.

We start with the analysis of Q and L. For this we introduce an arbitrary orientation on G (which is necessary for technical reasons).

Consider the $E \times X$ **edge-vertex matrix** ∇,

$$\nabla := (\nabla_{e,x})_{(e,x) \in E \times X}, \quad \nabla_{e,x} = \begin{cases} 1, & e \sim x, \\ 0, & \text{otherwise}, \end{cases}$$

then $\nabla : \mathbb{K}^X \to \mathbb{K}^E$ with $(\nabla f)(e) = f(\iota e) + f(\tau e)$. Now we calculate the $X \times X$ matrix $\nabla^* \nabla$:

$$(\nabla^* \nabla)_{x,y} = \sum_{e \in E} (\nabla^*)_{x,e} \nabla_{e,y} = \sum_{e \in E} \nabla_{e,x} \nabla_{e,y}.$$

For $x \neq y$ we have

$$\nabla_{e,x} \nabla_{e,y} = \begin{cases} 1, & e \sim x \text{ and } e \sim y, \\ 0, & \text{otherwise}, \end{cases} = \begin{cases} 1, & e = \{x, y\}, \\ 0, & \text{otherwise}, \end{cases}$$

therefore,

$$(\nabla^* \nabla)_{x,y} = \sum_{e : e \sim x \text{ and } e \sim y} 1 = \begin{cases} 1, & x \sim y, \\ 0, & \text{otherwise}. \end{cases}$$

For the diagonal elements we have

$$(\nabla^*\nabla)_{x,x} = \sum_{e:\, e\sim x} 1 = \deg x.$$

So overall we have

$$\nabla^*\nabla = D + A \equiv Q. \qquad (1.7)$$

A similar representation holds for L. Define the $E \times X$ **oriented edge-vertex matrix** $\vec{\nabla}$,

$$\vec{\nabla} := (\vec{\nabla}_{e,x})_{(e,x)\in E\times X}, \qquad \vec{\nabla}_{e,x} = \begin{cases} +1, & x = \iota e, \\ -1, & x = \tau e, \\ 0, & \text{otherwise.} \end{cases}$$

We have analogously $\vec{\nabla} : \mathbb{K}^X \to \mathbb{K}^E$ with

$$(\vec{\nabla} f)(e) = f(\iota e) - f(\tau e),$$

$$(\vec{\nabla}^*\vec{\nabla})_{x,y} = \sum_{e\in E} (\vec{\nabla}^*)_{x,e}\, \vec{\nabla}_{e,y} = \sum_{e\in E} \vec{\nabla}_{e,x}\, \vec{\nabla}_{e,y}.$$

For $x \neq y$ we have

$$\vec{\nabla}_{e,x}\, \vec{\nabla}_{e,y} = \begin{cases} -1, & e = \{x, y\}, \\ 0, & \text{otherwise,} \end{cases} \qquad (\vec{\nabla}^*\vec{\nabla})_{x,y} = \begin{cases} -1, & x \sim y, \\ 0, & \text{otherwise.} \end{cases}$$

Furthermore, for each $x \in X$ we have

$$\vec{\nabla}_{e,x}\, \vec{\nabla}_{e,x} = \begin{cases} (+1)\cdot(+1), & x = \iota e, \\ (-1)\cdot(-1), & x = \tau e, \\ 0, & \text{otherwise} \end{cases} = \begin{cases} 1, & e \sim x, \\ 0, & \text{otherwise.} \end{cases}$$

and

$$(\vec{\nabla}^*\vec{\nabla})_{x,x} = \sum_{e:\, e\sim x} 1 = \deg x.$$

Overall, we get

$$\vec{\nabla}^*\vec{\nabla} = D - A \equiv L. \qquad (1.8)$$

The two representations (1.7) and (1.8) will play an important role.

Remark 1.5.4 The transposed matrices ∇^* and $\vec{\nabla}^*$ are often referred to as the **incidence matrix** and the **oriented incidence matrix** of G, respectively.

Lemma 1.5.5 *For all $f \in \mathbb{K}^X$ the identities*

$$\langle f, Qf \rangle = \sum_{e \in E} \left| f(\iota e) + f(\tau e) \right|^2, \tag{1.9}$$

$$\langle f, Lf \rangle = \sum_{e \in E} \left| f(\iota e) - f(\tau e) \right|^2. \tag{1.10}$$

hold. In particular, all eigenvalues of L and Q are nonnegative.

Proof For all $f \in \mathbb{K}^X$ we have

$$\langle f, Qf \rangle = \langle f, \nabla^* \nabla f \rangle_{\mathbb{K}^X} = \langle \nabla f, \nabla f \rangle_{\mathbb{K}^E} = \| \nabla f \|^2_{\mathbb{K}^E}$$

$$= \sum_{e \in E} \left| (\nabla f)(e) \right|^2 = \sum_{e \in E} \left| f(\iota e) + f(\tau e) \right|^2 \geq 0,$$

$$\langle f, Lf \rangle = \langle f, \vec{\nabla}^* \vec{\nabla} f \rangle_{\mathbb{K}^X} = \langle \vec{\nabla} f, \vec{\nabla} f \rangle_{\mathbb{K}^E} = \| \vec{\nabla} f \|^2_{\mathbb{K}^E}$$

$$= \sum_{e \in E} \left| (\vec{\nabla} f)(e) \right|^2 = \sum_{e \in E} \left| f(\iota e) - f(\tau e) \right|^2 \geq 0.$$

If f is an eigenvector of Q with eigenvalue λ, then

$$0 \leq \langle f, Qf \rangle = \langle f, \lambda f \rangle = \lambda \langle f, f \rangle = \lambda \| f \|^2,$$

and since $\| f \| > 0$ we obtain $\lambda \geq 0$. The same argument also applies to the eigenvalues of L. $\qquad\square$

Remark 1.5.6 We will often use the following notation: For a subset $Y \subset X$, $\mathbb{1}_Y$ is the **indicator function of Y**, i.e. $\mathbb{1}_Y : X \to \mathbb{K}$,

$$\mathbb{1}_Y(x) = \begin{cases} 1, & x \in Y, \\ 0, & \text{otherwise,} \end{cases}$$

and we set

$$\mathbb{1} := \mathbb{1}_X.$$

Lemma 1.5.7 *If the graph G is <u>connected</u>, then $\ker L = \mathbb{K}\mathbb{1}$, in particular $\dim \ker L = 1$.*

Proof From the representation (1.3) it follows that $L\mathbb{1} = 0$, so $\mathbb{K}\mathbb{1} \subset \ker L$. If $f \in \ker L$, then $Lf = 0$ and $\langle f, Lf \rangle = 0$. From (1.10) it follows

$$f(\iota e) = f(\tau e) \text{ for all } e \in E. \tag{1.11}$$

Let $x, y \in X$ with $x \neq y$. Since G is connected, there is a path $(x = x_0, x_1, \ldots, x_k = y)$ from x to y. For each $j \in \{1, \ldots, k\}$ one has $\{x_{j-1}, x_j\} \in E$, and from (1.11) it follows that $f(x_{j-1}) = f(x_j)$. Then

$$f(x) \equiv f(x_0) = f(x_1) = \cdots = f(x_k) \equiv f(y).$$

Since x and y were chosen arbitrarily, f is constant, i.e. $f = c\mathbb{1}$ for some $c \in \mathbb{K}$. In other words, $f \in \mathbb{K}\mathbb{1}$ and thus $\ker L \subset \mathbb{K}\mathbb{1}$. $\qquad\square$

We can summarize all observations as follows:

Theorem 1.5.8 (Smallest Eigenvalue of L) *The smallest eigenvalue of L is always 0. If G is a graph with N connected components $G_j := (X_j, E_j)$, $j \in \{1, \ldots, N\}$, then*

$$\dim \ker L = N, \quad \ker L = \operatorname{span}\left\{ \mathbb{1}_{X_j} : j \in \{1, \ldots, N\} \right\}.$$

In particular, G is connected if and only if the smallest eigenvalue of L is simple.

Proof Lemma 1.5.5 shows that L has no negative eigenvalues.

Let L_j be the Laplace matrices of G_j, then $L = L_1 \oplus \cdots \oplus L_N$. By Lemma 1.5.7, $\ker L_j = \mathbb{K}\mathbb{1}_{X_j}$ and $\dim \ker L_j = 1$. Then

$$\dim \ker L = \sum_{j=1}^{N} \dim \ker L_j = \sum_{j=1}^{N} 1 = N,$$

and since $N \geq 1$, the value 0 is indeed an eigenvalue of L. Furthermore, we have

$$\ker L = \bigoplus_{j=1}^{N} \ker L_j = \bigoplus_{j=1}^{N} \mathbb{K}\mathbb{1}_{X_j} = \operatorname{span}\left\{ \mathbb{1}_{X_j} : j \in \{1, \ldots, N\} \right\}.$$

$\qquad\square$

Now we consider similar constructions for the matrix Q.

Lemma 1.5.9 *If G is a <u>connected</u> graph, then the following conditions are equivalent:*

(a) G is bipartite.

(b) There exists an $f \in \mathbb{K}^X$ with $f \not\equiv 0$ such that

$$f(\iota e) = -f(\tau e) \text{ for all } e \in E. \tag{1.12}$$

If one of these conditions is fulfilled, then

$$\dim \mathcal{F} = 1 \quad \text{for } \mathcal{F} := \left\{ f \in \mathbb{K}^X : f \text{ satisfies (1.12)} \right\}. \tag{1.13}$$

Proof (a) $\Rightarrow$(b)   If G is bipartite, then there exist disjoint $Y, Y' \subset X$ with $X = Y \cup Y'$ such that each edge goes from Y to Y'. The function $f := \mathbb{1}_Y - \mathbb{1}_{Y'}$, i.e.

$$f : X \ni x \mapsto \begin{cases} 1, & x \in Y, \\ -1, & x \in Y', \end{cases}$$

obviously satisfies condition (1.12).

(b) $\Rightarrow$(a)   Let f be a function with the required properties. Then there exists an $x \in X$ with $c := f(x) \neq 0$. Now let $y \in X$ be any vertex with $x \neq y$, then there is a path $(x = x_0, x_1, \ldots, x_k = y)$ from x to y. For each $j \in \{1, \ldots, k\}$, $\{x_{j-1}, x_j\} \in E$ holds, therefore $f(x_j) = -f(x_{j-1})$ thanks to (1.12) and then $f(x_j) = (-1)^j f(x_0)$. It follows

$$f(y) \equiv f(x_k) = (-1)^k f(x_0) \equiv (-1)^k f(x) = (-1)^k c.$$

Thus, $f(y) \in \{c, -c\}$ holds for every $y \in X$. The sets

$$Y := \{y \in X : f(y) = c\}, \quad Y' := \{y \in X : f(y) = -c\}$$

are obviously disjoint and thanks to (1.12) there is no edge from Y to Y or from Y' to Y', so G is bipartite.

This proves the equivalence (a) $\Leftrightarrow$(b).

Now let (a) be fulfilled. If G has no edges, then $|X| = 1$ (since G is connected) and $\mathcal{F} = \mathbb{K}^X = \mathbb{K}$, so $\dim \mathcal{F} = 1$. Assume now that G has at least one edge $\{y, y'\}$. Then there are disjoint Y and Y' with $X = Y \cup Y'$ such that every edge goes from Y to Y', and we can additionally assume $y \in Y$ and $y' \in Y'$.

Let $x \in Y$ with $x \neq y$, then every path $(x = x_0, \ldots, x_k = y)$ from x to y has an <u>even</u> length (since with each step one switches from Y to Y' or from Y' to Y), i.e. $k \in 2\mathbb{N}$. For each $j \in \{1, \ldots, k\}$ one has $f(x_j) = -f(x_{j-1})$, so

$$f(y) \equiv f(x_k) = (-1)^k f(x_0) = f(x_0) \equiv f(x),$$

i.e., f is constant on Y. The same argument shows that f is also constant on Y', so there are $c, c' \in \mathbb{K}$ with $f := c\mathbb{1}_Y + c'\mathbb{1}_{Y'}$. Since y and y' are connected by an edge, it holds

$$c' \equiv f(y') = -f(y) \equiv -c,$$

and overall we get $f = c(\mathbb{1}_Y - \mathbb{1}_{Y'})$. With this we have shown $\mathcal{F} = \mathbb{K}(\mathbb{1}_Y - \mathbb{1}_{Y'})$ and consequently $\dim \mathcal{F} = 1$. $\qquad\square$

Corollary 1.5.10 *If G is <u>connected</u>, then*

$$\dim \ker Q = \begin{cases} 1, & G \text{ is bipartite}, \\ 0, & \text{otherwise.} \end{cases}$$

In particular, the following conditions are equivalent:

(a) G is bipartite.
(b) 0 is an eigenvalue of Q.
(c) The smallest eigenvalue of Q is 0.

Proof Let $f \in \ker Q$, then $Qf = 0$ and, with the help of (1.9),

$$0 = \langle f, Qf \rangle = \sum_{e \in E} \left| f(\iota e) + f(\tau e) \right|^2,$$

i.e., f satisfies (1.12). Conversely, if f satisfies condition (1.12), it immediately follows with (1.4), that $Qf = 0$. With this we have shown

$$\ker Q = \left\{ f \in \mathbb{K}^X : f \text{ satisfies (1.12)} \right\},$$

and one uses the previous lemma 1.5.9. $\qquad\square$

From these considerations, the following theorem arises:

Theorem 1.5.11 (Smallest Eigenvalue of Q) *The matrix Q has 0 as an eigenvalue exactly when Q has a bipartite connected component. More precisely,*

$$\dim \ker Q = \text{the number of bipartite connected components of } G.$$

In particular, Q has at least one bipartite connected component exactly when the smallest eigenvalue of Q is 0.

Proof Let $G_1, \ldots, G_N$ be the connected components of G and Q_j the corresponding unsigned Laplace matrices. Then

$$Q = \bigoplus_{j=1}^{N} Q_j, \quad \dim \ker Q = \sum_{j=1}^{N} \dim \ker Q_j,$$

and one uses Corollary 1.5.10 for each j. Since Q has no negative eigenvalues (Lemma 1.5.5), the last statement also holds. $\qquad\square$

In addition, the following statement is often used:

Theorem 1.5.12 (Trace Theorems for L and Q) *Let L and Q be the Laplace matrix and the unsigned Laplace matrix of a graph $G = (X, E)$ with n vertices and let*

- *$\lambda_1, \ldots, \lambda_n$ be the eigenvalues of L,*
- *$\mu_1, \ldots, \mu_n$ be the eigenvalues of Q,*

then it holds

$$\lambda_1 + \cdots + \lambda_n = \mu_1 + \cdots + \mu_n = 2|E|.$$

Proof According to the general trace theorem (Remark 1.4.3), the sum of the eigenvalues is equal to the trace of the corresponding matrix. For each $x \in X$, $L_{x,x} = Q_{x,x} = \deg x$, therefore, according to the handshake lemma (Lemma 1.1.4):

$$\operatorname{tr} L = \operatorname{tr} Q = \sum_{x \in X} \deg x = 2|E|.$$

$\qquad\square$

Now we consider the first spectral properties of the adjacency matrix A. The following property will be used several times.

Theorem 1.5.13 (Geometric Meaning of A^m) *Let $m \in \mathbb{N}$ and $x, y \in X$, then*

$$(A^m)_{x,y} = \text{the number of walks } \underline{\text{of length } m} \text{ from } x \text{ to } y.$$

Proof Denote

$$s_{x,y}^m := \text{the number of walks of length } m \text{ from } x \text{ to } y.$$

We have to prove that $(A^m)_{x,y} = s_{x,y}^m$ for all $m \in \mathbb{N}$ and all $x, y \in X$. For this, induction over m is used.

For $m = 1$ the statement is true: The number $s_{x,y}^1$ of walks of length 1 from x to y is 1 for $x \sim y$ and 0 otherwise, i.e., it is equal to $A_{x,y}$.

Assume that the statement holds for all $m \le M$ (for a $M \in \mathbb{N}$). Every walk $(x = x_0, \ldots, x_M, x_{M+1} = y)$ of length $M + 1$ from x to y looks as follows: x_M is a neighbor of y and $(x_0, \ldots, x_M)$ is a walk of length M from x to x_M. It follows by summing up

$$s_{x,y}^{M+1} = \sum_{z:\, z \sim y} s_{x,z}^M.$$

According to the assumption, $s_{x,z}^M = (A^M)_{x,z}$ holds for all $x, z \in X$, thus

$$s_{x,y}^{M+1} = \sum_{z:\, z \sim y} (A^M)_{x,z} = \sum_{z \in X} (A^M)_{x,z} A_{z,y} = (A^{M+1})_{x,y}.$$

$\square$

Corollary 1.5.14 (Trace Theorem for the Adjacency Matrix) *Let* $\lambda_1, \ldots, \lambda_n$ *be the eigenvalues of A. Then the following identities hold:*

$$\sum_{j=1}^{n} \lambda_j = 0, \qquad \sum_{j=1}^{n} \lambda_j^2 = 2|E|,$$

$$\sum_{j=1}^{n} \lambda_j^3 = 6 \times \text{ the number of triangles in } G.$$

Proof For each $m \in \mathbb{N}$ the numbers $\lambda_1^m, \ldots, \lambda_n^m$ are the eigenvalues of A^m. According to the trace theorem (Remark 1.4.3), it holds

$$\sum_{j=1}^{n} \lambda_j^m = \text{tr}(A^m) \equiv \sum_{x \in X} (A^m)_{x,x}.$$

For $m = 1$ we obtain the first desired formula, since $A_{x,x} = 0$ for all $x \in X$. Every walk of length 2 from x to x has the form (x, y, x) with $y \sim x$. It follows (with $m = 2$)

$$\sum_{j=1}^{n} \lambda_j^2 = \sum_{x \in X} \text{ the number of walks of length 2 from } x \text{ to } x$$

$$= \sum_{x \in X} \sum_{y:\, y \sim x} 1 = \sum_{x \in X} \deg x = 2|E|,$$

where the last equality holds by the handshake lemma (Lemma 1.1.4).

Each triangle can be represented in 6 different ways as a closed walk of length 3 (one can choose the starting vertex and the direction of traversal). Therefore, for $m = 3$:

$$\sum_{j=1}^{n} \lambda_j^3 = \sum_{x \in X} \text{the number of walks of length 3 from } x \text{ to } x$$

$$= 6 \times \text{the number of triangles in } G.$$

$\square$

Exercise 1.5.1

1. Let M be a $n \times n$ matrix, whose coefficients are integers, and let λ be an eigenvalue of M with $\lambda \in \mathbb{Q}$. We want to show that $\lambda \in \mathbb{Z}$.
   (a) Show that an eigenvector with rational coefficients exists.
       Hint: Think of Gaussian elimination.
   (b) Show: There exists an eigenvector $x = (x_1, \ldots, x_n)$ with $x_i \in \mathbb{Z}$ and $\gcd(x_1, \ldots, x_n) = 1$.
   (c) Deduce that $\lambda \in \mathbb{Z}$.
2. Are there graphs with $\frac{1}{2} \in \sigma(A)$? with $\frac{3}{4} \in \sigma(L)$?

Exercise 1.5.2 Let $G = (X, E)$ be a connected graph. Let L be the corresponding Laplace matrix and λ_i its eigenvalues.

1. Determine the eigenvalues of $L + cJ$ with $c \in \mathbb{R}$. Here and further, J is the matrix, all of whose elements are equal to 1.
   Hint: Test on an orthonormal basis of eigenvectors of L.
2. Determine the L-eigenvalues of the complement graph $\overline{G}$ of G.

Exercise 1.5.3 Let $G = (X, E)$ be a connected graph and A its adjacency matrix. The *diameter* $d(G)$ of G is defined by

$$d(G) = \max_{u,v \in X,\, u \neq v} d(u, v),$$

where $d(u, v)$ is the distance between u and v.

Let m be the number of pairwise different eigenvalues of A. We want to show:

$$m \geq d(G) + 1.$$

1. Show: A^m is a linear combination of the A^k with $k \in \{0, \ldots, m - 1\}$.
2. Assume that $m \leq d(G)$.
   (a) Show that $A^{d(G)}$ is a linear combination of A^k with $k \leq d(G) - 1$.
   (b) Let $u, v \in X$ with $d(u, v) = d(G)$. What can be said about $(A^k)_{uv}$ with $k \leq d(G) - 1$ and $k = d(G)$?
   (c) Find a contradiction.

1.6   Important Graphs

We now consider some specific classes of graphs that are extensively used in the book, and for which the eigenvalues of the most important graph matrices can be explicitly calculated.

Example 1.6.1 A graph $G := (X, E)$ is called **regular**, if all vertices have the same degree: There is a $k \in \mathbb{N}_0$, such that $\deg x = k$ for all $x \in X$ (in this case, we say that G is **regular of degree** k or k**-regular**). In this case, $D = kI$ and $L := D - A = kI - A$. If $\lambda_1, \ldots, \lambda_n$ are the eigenvalues of A, then $k - \lambda_1, \ldots, k - \lambda_n$ are the eigenvalues of L.

Example 1.6.2 For $n \in \mathbb{N}$, K_n denotes the **complete graph** with n vertices, i.e. $K_n := (X, E)$ with $|X| = n$ and $x \sim y$ for all $x, y \in X$ with $x \neq y$ (any two vertices are connected by an edge), see Fig. 1.16. This is obviously a regular graph of degree $n - 1$. The corresponding adjacency matrix is

$$A = (A_{x,y}), \quad A_{x,y} = \begin{cases} 1, & x \neq y, \\ 0, & x = y. \end{cases}$$

If we define

$$I_n := \text{the } n \times n \text{ identity matrix,}$$

$$J_n := \text{the } n \times n \text{ ones matrix (all coefficients are equal to 1)}$$

then $A = J_n - I_n$ and $L = D - A = (n - 1)I_n - (J_n - I_n) = nI_n - J_n$.

The spectrum of J_n will play a role in many places. First, we note that J_n is real and symmetric and thus defines a self-adjoint linear operator in $\mathbb{K}^n$. Furthermore, $J_n \mathbb{1} = n\mathbb{1}$. Let now $f \in \mathbb{K}^n$ with $f \perp \mathbb{1}$. Then

$$J_n f = \begin{pmatrix} \langle \mathbb{1}, f \rangle \\ \ldots \\ \langle \mathbb{1}, f \rangle \end{pmatrix} = 0,$$

i.e. $\mathbb{1}^\perp \subset \ker J_n$. Since $\dim \mathbb{1}^\perp = n - 1$, any orthonormal basis $(b_1, \ldots, b_{n-1})$ of it can be completed with $\frac{1}{\sqrt{n}}\mathbb{1}$ to an orthonormal basis of $\mathbb{K}^n$. From $J_n \mathbb{1} = n\mathbb{1}$ and $J_n b_j = 0$ it follows

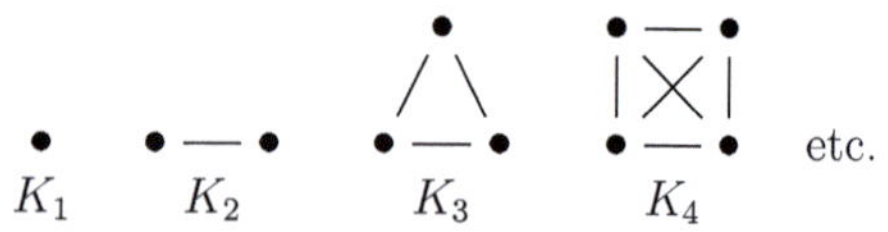

Fig. 1.16 Complete graphs

$$K_1 \qquad K_2 \qquad K_3 \qquad K_4 \qquad \text{etc.}$$

$$\sigma(J_n) = \{\underbrace{0, \ldots, 0}_{n-1 \text{ times}}, n\}.$$

This implies

$$\sigma(A) = \{\underbrace{-1, \ldots, -1}_{n-1 \text{ times}}, n - 1\}, \qquad \sigma(L) = \{\underbrace{n, \ldots, n}_{n-1 \text{ times}}, 0\}.$$

Now we consider the spectra of bipartite graphs.

Proposition 1.6.3 *Let A be the adjacency matrix of a <u>bipartite</u> graph, then*

$$\sigma(A) = -\sigma(A), \tag{1.14}$$

or more precisely, $\dim \ker(A - \lambda I) = \dim \ker(A + \lambda I)$ *for all $\lambda \in \mathbb{R}$.*

Proof Let G be bipartite, $X = Y \cup Y'$ for disjoint Y and Y', such that all edges are between Y and Y'. Then A is a block matrix,

$$A = \begin{pmatrix} 0 & B \\ B^* & 0 \end{pmatrix},$$

where B is a $Y \times Y'$ matrix. Consider the block matrix

$$\Theta := \begin{pmatrix} I & 0 \\ 0 & -I \end{pmatrix}$$

(with the blocks of the same size as in A). Obviously, Θ is invertible, $\Theta^{-1} = \Theta$, and

$$\Theta^{-1} A \Theta \equiv \Theta A \Theta = \begin{pmatrix} I & 0 \\ 0 & -I \end{pmatrix} \begin{pmatrix} 0 & B \\ B^* & 0 \end{pmatrix} \begin{pmatrix} I & 0 \\ 0 & -I \end{pmatrix} = \begin{pmatrix} 0 & -B \\ -B^* & 0 \end{pmatrix} = -A.$$

It follows $\Theta(-A) = A\Theta$, so (Proposition 1.4.6) $\sigma(A) = \sigma(-A) = -\sigma(A)$. $\qquad \square$

Remark 1.6.4 Later we will see (Sect. 2.5) that the converse also holds: If the adjacency matrix satisfies (1.14), the graph is bipartite.

For the practical calculation of the eigenvalues of the adjacency matrix for bipartite graphs, the following rule is important:

Lemma 1.6.5 *Let B be a $m \times n$ matrix and C be a $n \times m$ matrix, then*

$$\dim \ker(BC - \lambda I) = \dim \ker(CB - \lambda I) \text{ for all } \lambda \in \mathbb{K} \setminus \{0\}. \tag{1.15}$$

In other words, the $m \times m$ matrix BC and the $n \times n$ matrix CB have the same eigenvalues (with the same multiplicities) outside of 0.

Proof Let $\lambda \in \mathbb{K} \setminus \{0\}$. For each $v \in \ker(CB - \lambda I)$ we have

$$CBv = \lambda v. \tag{1.16}$$

If $v \neq 0$, then $Bv \neq 0$. This shows that the restriction of B to $\ker(CB - \lambda I)$ is injective. Now we apply B to both sides of (1.16) and get $B(CBv) = B(\lambda v)$. This can be rewritten as $(BC)Bv = \lambda Bv$ and it follows $Bv \in \ker(BC - \lambda I)$.

From these considerations it follows that $B : \ker(CB - \lambda I) \to \ker(BC - \lambda I)$ is an injective linear mapping, in particular

$$\dim \ker(CB - \lambda I) \leq \dim \ker(BC - \lambda I).$$

We can swap the roles of B and C, then we also get the reverse inequality. Thus, the two dimensions are equal and (1.15) follows. $\square$

Theorem 1.6.6 *Let A be the adjacency matrix of a bipartite graph,*

$$A = \begin{pmatrix} 0 & B \\ B^* & 0 \end{pmatrix}$$

with a $m \times n$ matrix B. If $\mu_1, \ldots, \mu_k$ are the distinct positive eigenvalues of BB^* with multiplicities $n_1, \ldots, n_k$, then

$$\sigma(A) = \Big\{ \underbrace{+\sqrt{\mu_1}}_{n_1 \text{ times}}, \underbrace{-\sqrt{\mu_1}}_{n_1 \text{ times}}, \ldots, \underbrace{+\sqrt{\mu_k}}_{n_k \text{ times}}, \underbrace{-\sqrt{\mu_k}}_{n_k \text{ times}}, \underbrace{0}_{n+m-2(n_1+\cdots+n_k) \text{ times}} \Big\}.$$

Proof We have $\sigma(A^2) = \sigma(A)^2$, see Remark 1.4.15, where

$$A^2 = \begin{pmatrix} 0 & B \\ B^* & 0 \end{pmatrix} \begin{pmatrix} 0 & B \\ B^* & 0 \end{pmatrix} = \begin{pmatrix} BB^* & 0 \\ 0 & B^*B \end{pmatrix}$$

so $\sigma(A^2) = \sigma(BB^*) \cup \sigma(B^*B)$. Thanks to Lemma 1.6.5 we get

$$\big\{ \lambda^2 : \lambda \in \sigma(A) \big\} = \Big\{ \underbrace{\mu_1}_{2n_1 \text{ times}}, \ldots, \underbrace{\mu_k}_{2n_k \text{ times}}, \underbrace{0}_{n+m-2(n_1+\cdots+n_k) \text{ times}} \Big\}.$$

This immediately gives us that 0 is an eigenvalue of A of multiplicity $n+m-2(n_1+\cdots+n_k)$. If $\lambda \in \sigma(A)$ with $\lambda \neq 0$, then λ is necessarily one of the numbers $\pm \sqrt{\mu_j}$. According to Proposition 1.6.3, $+\sqrt{\mu_j}$ and $-\sqrt{\mu_j}$ have the same multiplicity in $\sigma(A)$, so the claim follows. $\square$

Fig. 1.17 A complete
bipartite graph

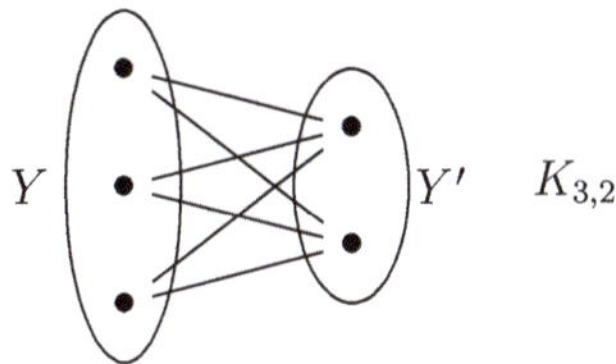

Example 1.6.7 Let $m, n \in \mathbb{N}$. The **complete bipartite graph** $K_{m,n}$ is defined as follows: $K_{m,n} = (X, E)$ with $X = Y \cup Y'$ and disjoint Y and Y' such that $|Y| = m$ and $|Y'| = n$, and

$$E = \big\{ \{y, y'\} : y \in Y,\ y' \in Y' \big\},$$

i.e., there is an edge between every vertex in Y and every vertex in Y' and no other edges (Fig. 1.17). The graphs $K_{m,n}$ and $K_{n,m}$ are obviously isomorphic.

The adjacency matrix of $K_{m,n}$ is

$$A = \begin{pmatrix} 0 & J_{m,n} \\ J_{n,m} & 0 \end{pmatrix},$$

where $J_{m,n} :=$ the $m \times n$ ones matrix (all coefficients are equal to 1).

We have $J_{m,n} J_{m,n}^* = J_{m,n} J_{n,m} = n J_m$, $\sigma(J_{m,n} J_{m,n}^*) = \sigma(n J_m) = \{\underbrace{0, \ldots, 0}_{m-1 \text{ times}}, mn\}$.

From Theorem 1.6.6 it follows

$$\sigma(A) = \big\{ -\sqrt{mn},\ \underbrace{0, \ldots, 0}_{m+n-2 \text{ times}},\ \sqrt{mn} \big\} \text{ for } K_{m,n}. \tag{1.17}$$

The equality (1.17) can also be proven differently. If one subtracts the first row of A from rows 2 to m and then subtracts the $(m + 1)$-th row from rows $m + 2$ to $m + n$, one obtains a matrix in which exactly $m + n - 2$ rows are zero. From the Gaussian elimination process, it follows that $\dim \ker A = m + n - 2$. The two remaining eigenvalues are $\pm \lambda$ with $\lambda > 0$. To determine λ, we consider the vectors

$$x = (\underbrace{a, \ldots, a}_{m \text{ times}}, \underbrace{b, \ldots, b}_{n \text{ times}})^*, \quad a, b \in \mathbb{K} \setminus \{0\},$$

then

$$Ax = (\underbrace{nb, \ldots, nb}_{m \text{ times}}, \underbrace{ma, \ldots, ma}_{n \text{ times}})^*.$$

For Ax to be a multiple of x, we must have $\frac{nb}{a} = \frac{ma}{b}$. It follows that $\left(\frac{b}{a}\right)^2 = \frac{m}{n}$ and for $b = \pm\sqrt{\frac{m}{n}}a$ we get $Ax = \pm\sqrt{mn}x$, i.e., the two missing eigenvalues are $\pm\sqrt{mn}$.

Now we calculate the eigenvalues of L for $K_{m,n}$. For $m = n$ the graph is n-regular, so

$$\sigma(L) = n - \sigma(A) = \{0, \underbrace{n, \ldots, n}_{2n-2 \text{ times}}, 2n\} \text{ for } K_{n,n}. \tag{1.18}$$

Assume that $m \neq n$, then

$$L = \begin{pmatrix} nI_m & -J_{m,n} \\ -J_{n,m} & mI_n \end{pmatrix}.$$

It holds

$$L - nI = \begin{pmatrix} 0 & -J_{m,n} \\ -J_{n,m} & (m-n)I_n \end{pmatrix}.$$

Since the first m rows are identical and the last n rows are linearly independent, it follows that $\dim\ker(L - nI) = m - 1$. Similarly, $\dim(L - mI) = n - 1$, In this case, 0 is a simple eigenvalue (since $K_{m,n}$ is obviously connected). With this, $m + n - 1$ eigenvalues are already determined: the simple eigenvalue 0, the $(m-1)$-fold eigenvalue n and the $(n-1)$-fold eigenvalue m. Let λ be the remaining eigenvalue. From the trace theorem for L (Theorem 1.5.12) it follows

$$\operatorname{tr} L \equiv 0 + (m-1)n + (n-1)m + \lambda = 2|E| = 2mn,$$

so $\lambda = m + n$. In total we have

$$\sigma(L) = \{0, \underbrace{n, \ldots, n}_{(m-1) \text{ times}}, \underbrace{m, \ldots, m}_{(n-1) \text{ times}}, m+n\} \text{ for } K_{m,n} \text{ with } m \neq n.$$

For $m = n$ the result agrees with (1.18). Similarly, one calculates the eigenvalues of Q and obtains $\sigma(Q) = \sigma(L)$.

Example 1.6.8 (Graphs with Symmetries)   A **symmetry** of a graph $G = (X, E)$ is a bijective mapping $\varphi : X \to X$ such that for all $x, y \in X$ it holds

$$\varphi(x) \sim \varphi(y) \text{ if and only if } x \sim y.$$

Let $R \in \{A, L, Q\}$, then $R_{\varphi(x),\varphi(y)} = R_{x,y}$ for all $x, y \in X$. We consider the bijective linear mapping

$$\Theta : \mathbb{K}^X \to \mathbb{K}^X, \quad (\Theta f)(x) := f\big(\varphi(x)\big) \text{ for all } x \in X.$$

The same calculation as in Remark 1.5.2 (Independence of the spectrum from graph isomorphisms) shows $\Theta R = R\Theta$.

Proposition 1.6.9 (Commuting Linear Mappings) *Let V be a vector space and $R, \Theta : V \to V$ commuting linear mappings, i.e. $\Theta R = R\Theta$. Then $\ker(\Theta - \mu I)$ for each $\mu \in \mathbb{K}$ is an invariant subspace of R.*

Proof Let $v \in \ker(\Theta - \mu I)$, then $\Theta v = \mu v$ and $\Theta(Rv) = R(\Theta v) = R(\mu v) = \mu R v$, i.e. $Rv \in \ker(\Theta - \mu I)$. $\square$

Remark 1.6.10 Proposition 1.6.9 is particularly useful when V can be represented as a direct sum of the eigenspaces $V_j := \ker(\Theta - \mu_j I)$. Then it is sufficient to determine the eigenvalues of $R : V_j \to V_j$. This is especially interesting when the dimensions of V_j are small. If for some μ the eigenspace $\ker(\Theta - \mu I)$ is one-dimensional, $\ker(\Theta - \mu I) = \mathbb{K}v$, then v is automatically an eigenvector of R: Since $Rv \in \ker(\Theta - \mu I) = \mathbb{K}v$, then there is a $\lambda \in \mathbb{K}$ with $Rv = \lambda v$.

If $\dim V = n$ and Θ has exactly n different eigenvalues $\mu_1, \dots, \mu_n \in \mathbb{K}$, then all subspaces $\ker(\Theta - \mu_j I)$ are one-dimensional, $\ker(\Theta - \mu_j I) = \mathbb{K}v_j$. Then the vectors $v_1, \dots, v_n$ are linearly independent and form a basis of V. Each v_j is then automatically an eigenvector of R, $Rv_j = \lambda_j v_j$ for suitable $\lambda_j \in \mathbb{K}$, and the numbers $\lambda_1, \dots, \lambda_n$ exhaust all eigenvalues of R.

Example 1.6.11 (Eigenvalues for the Cycle Graphs)   Let C_n be a cycle graph with n vertices ($n \geq 3$), see Fig. 1.18. More precisely, $C_n = (X, E)$,

$$X := \{0, 1, \dots, n - 1\}, \qquad E := \Big\{\{x, \, x + 1 \bmod n\} : \, x \in X\Big\}.$$

The graph is obviously invariant with respect to cyclic shifts, i.e., the mapping

$$\varphi : X \to X, \quad \varphi(x) = x + 1 \bmod n,$$

is a symmetry of C_n. Consider the associated mapping

$$\Theta_n : \mathbb{K}^X \to \mathbb{K}^X, \quad (\Theta_n f)(x) = f(x + 1 \bmod n).$$

Fig. 1.18 The cycle graph C_5

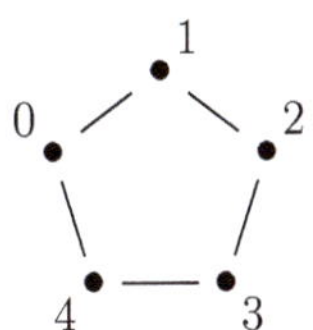

It holds that $\Theta_n^n = I_n$, so every eigenvalue μ of Θ satisfies the equation $\mu^n = 1$. Conversely, if $\mu \in \mathbb{K}$ with $\mu^n = 1$, then $f : X \ni x \mapsto \mu^x \in \mathbb{K}$ is an eigenvector of Θ_n to μ.

In this case, it is convenient to consider $\mathbb{K} := \mathbb{C}$. Then $\sigma(\Theta_n)$ consists of the n roots of unity

$$\mu_k := e^{\frac{2\pi i k}{n}}, \quad k \in \{0, \ldots, n-1\}, \tag{1.19}$$

and the corresponding eigenspaces $\ker(\Theta_n - \mu_k I)$ are one-dimensional and spanned by

$$f_k : X \ni x \mapsto \mu_k^x \in \mathbb{C}$$

As discussed in Remark 1.6.10, each f_k is automatically an eigenvector of the adjacency matrix A for C_n. For each $f \in \mathbb{C}^X$ it holds

$$(Af)(x) = f(x - 1 \bmod n) + f(x + 1 \bmod n), \quad x \in X,$$

and for $f := f_k$ we obtain

$$(Af_k)(x) = \mu_k^{x-1} + \mu_k^{x+1} = (\mu_k^{-1} + \mu_k)\mu_k^x \equiv \left(2\cos\frac{2\pi k}{n}\right) f_k(x).$$

It follows that

$$\sigma(A) = \left\{2\cos\frac{2\pi k}{n} \; : \; k \in \{0, 1, \ldots, n-1\}\right\}.$$

The graph C_n is 2-regular, so $L = 2I - A$ and $Q = 2I + A$,

$$\sigma(L) = \left\{2 - 2\cos\frac{2\pi k}{n} \; : \; k \in \{0, 1, \ldots, n-1\}\right\},$$

$$\sigma(Q) = \left\{2 + 2\cos\frac{2\pi k}{n} \; : \; k \in \{0, 1, \ldots, n-1\}\right\}.$$

Example 1.6.12 Let B_n be the graph with $2n$ vertices, which is created by connecting each vertex of C_n with a new vertex (Fig. 1.19). The formal description is

$$B_n = (X, E), \quad X = \{x_0, x_1, \ldots, x_{n-1}, y_0, y_1, \ldots, y_{n-1}\},$$

$$E = \left\{\{x_j, x_{j+1 \bmod n}\} : \; j \in \{0, \ldots, n-1\}\right\} \cup \left\{\{x_j, y_j\} : \; j \in \{0, \ldots, n-1\}\right\}.$$

The graph B_n is less classical than the previous graphs, but we want to illustrate the method with symmetries once again. It will again be convenient to set $\mathbb{K} := \mathbb{C}$, and we only consider the Laplace matrix L. In this case the Laplace matrix L is given by

Fig. 1.19 The graph B_5

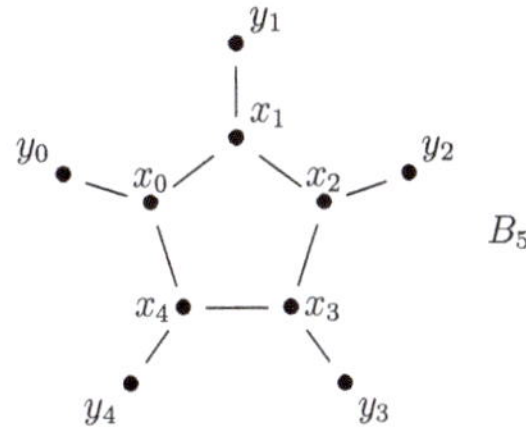

$$(Lf)(x_j) = 3f(x_j) - f(x_{j-1 \bmod n}) - f(x_{j+1 \bmod n}) - f(y_j),$$
$$(Lf)(y_j) = f(y_j) - f(x_j). \tag{1.20}$$

The graph has the symmetry

$$\varphi(x_j) = x_{j+1 \bmod n}, \quad \varphi(y_j) = y_{j+1 \bmod n}, \quad j \in \{0, 1, \ldots, n-1\},$$

which then generates the corresponding mapping $\Theta : \mathbb{C}^X \to \mathbb{C}^X$,

$$(\Theta f)(x_j) = f(x_{j+1 \bmod n}), \quad (\Theta f)(y_j) = f(y_{j+1 \bmod n}), \quad j \in \{0, 1, \ldots, n-1\},$$

i.e. $\Theta = \Theta_n \oplus \Theta_n$ with Θ_n from Example 1.6.11. By construction, Θ commutes with L. The eigenvalues of Θ are the same as for Θ_n, i.e. the numbers μ_k from (1.19), but with double multiplicity:

$$V_k := \ker\left(\Theta - \mu_k I\right) = \Big\{ f \in \mathbb{C}^X, \ f(x_j) = a\mu_k^j, \ f(y_j) = b\mu_k^j :$$
$$j \in \{0, \ldots, n-1\}, \ a, b \in \mathbb{C} \Big\}. \tag{1.21}$$

Each V_k is a two-dimensional invariant subspace of L. Moreover, $V_0, \ldots V_{n-1}$ are linearly independent and thus generate the entire vector space $\mathbb{C}^X$. Therefore, it is sufficient to determine the eigenvalues of $L : V_k \to V_k$ for each k.

For this, we will substitute the representation of $f \in V_k$ with $a, b \in \mathbb{C}$ as in (1.21) into the formulas (1.20). The condition $Lf = \lambda f$ takes the form

$$3a\mu_k^j - a\mu_k^{j-1} - a\mu_k^{j+1} - b\mu_k^j = \lambda a\mu_k^j, \qquad b\mu_k^j - a\mu_k^j = \lambda b\mu_k^j,$$

i.e.

$$\left(3 - 2\cos\tfrac{2\pi k}{n} - \lambda\right)a - b = 0,$$
$$-a + (1 - \lambda)b = 0.$$

From $f \neq 0$ it follows that $(a, b) \neq (0, 0)$, so the determinant of the system must be zero,

$$\det \begin{pmatrix} 3 - 2\cos\frac{2\pi k}{n} - \lambda & -1 \\ -1 & 1 - \lambda \end{pmatrix} = 0,$$

$$\text{i.e. } \lambda^2 - 2\left(2 - \cos\frac{2\pi k}{n}\right)\lambda + \left(2 - 2\cos\frac{2\pi k}{n}\right) = 0.$$

We obtain

$$\lambda = 2 - \cos\frac{2\pi k}{n} \pm \sqrt{\left(2 - \cos\frac{2\pi k}{n}\right)^2 - \left(2 - 2\cos\frac{2\pi k}{n}\right)}$$

$$= 2 - \cos\frac{2\pi k}{n} \pm \sqrt{2 - 2\cos\frac{2\pi k}{n} + \cos^2\frac{2\pi k}{n}}$$

and

$$\sigma(L) := \left\{ 2 - \cos\frac{2\pi k}{n} + \sqrt{2 - 2\cos\frac{2\pi k}{n} + \cos^2\frac{2\pi k}{n}} : k \in \{0, 1, \ldots, n - 1\} \right\}$$

$$\cup \left\{ 2 - \cos\frac{2\pi k}{n} - \sqrt{2 - 2\cos\frac{2\pi k}{n} + \cos^2\frac{2\pi k}{n}} : k \in \{0, 1, \ldots, n - 1\} \right\}.$$

Remark 1.6.13 Additional techniques for determining or estimating the eigenvalues of graph matrices (in closed form or by numerical algorithms) can be found, for example, in the book by Hoppen et al. [66].

Exercise 1.6.1 Let G be a graph and λ_i its Q-eigenvalues. Show that the A-spectrum of the line graph K_G consists of the eigenvalues $\lambda_i - 2$ with $\lambda_i \neq 0$ and the eigenvalue -2 (with corresponding multiplicities).

Hint: Let ∇ be the edge-vertex matrix of G. Consider $\nabla\nabla^*$ and $\nabla^*\nabla$.

Exercise 1.6.2 Let $G = (X, E)$ be a bipartite graph and let K_G be its line graph. Let λ_i be the L-eigenvalues of G.

1. Show: There exists an orientation and the corresponding oriented edge-vertex matrix $\vec{\nabla}$ for G, such that $\vec{\nabla}\,\vec{\nabla}^* - 2I$ is the adjacency matrix of K_G.
2. Show that the A-spectrum of K_G consists of the eigenvalues $\lambda_i - 2$ with $\lambda_i \neq 0$ and the eigenvalue -2 (with corresponding multiplicity).

Exercise 1.6.3 Let $G = (X, E)$ and $G' = (X', E')$ be graphs. Define a new graph $G\,\square\,G'$:

$$G\,\square\,G' := (X \times X', \widetilde{E}) \text{ with } \left\{(x, x'), (y, y')\right\} \in \widetilde{E} \iff \begin{cases} \left(x = y \text{ and } \{x', y'\} \in E' \right) \\ \text{or} \\ \left(\{x, y\} \in E \text{ and } x' = y' \right). \end{cases}$$

1. Show: $\sigma(A_{G \,\square\, G'}) = \sigma(A_G) + \sigma(A_{G'})$.

   Note: Let (f_j) and $(f'_{j'})$ be eigenbases for A_G and $A_{G'}$. Consider the functions

$$g_{j,j'} : \; X \times X' \ni (x, x') \mapsto f_j(x) f'_{j'}(x').$$

2. For $n \in \mathbb{N}$, let $H_n := \underbrace{K_2 \,\square\, \cdots \,\square\, K_2}_{n\text{-times}}$.

   (a) Draw H_n for $n \in \{2, 3\}$.

   (b) Determine the A-eigenvalues of H_n.

3. Show that for a suitable numbering of the vertices of H_n the following recursive formula holds:

$$A_{H_1} := \begin{pmatrix} 0 & 1 \\ 1 & 0 \end{pmatrix}, \quad A_{H_n} = \begin{pmatrix} A_{H_{n-1}} & I \\ I & A_{H_{n-1}} \end{pmatrix} \quad \text{for all } n \geq 2.$$

Exercise 1.6.4 Let $W_n = (X, E)$ be the path graph with n vertices ($n \in \mathbb{N}$ with $n \geq 2$) as illustrated in Fig. 1.20:

$$X = \{1, \ldots, n\}, \quad E = \big\{\{k, k+1\} : k \in \{1, \ldots, n-1\}\big\}.$$

We will calculate the A- and L-eigenvalues of W_n by transferring the calculations for the cycle C_m (Example 1.6.11) to W_n.

1. Let $f : \{0, \ldots, m-1\} \to \mathbb{K}$ be an eigenfunction of A_{C_m} or L_{C_m}. Show: There is <u>no</u> $k \in \{0, \ldots, m-1\}$ with

$$f(k) = f(k+1 \bmod m) = 0.$$

2. Let $f : \{0, \ldots, 2n+1\} \to \mathbb{K}$ be an eigenfunction of $A_{C_{2n+2}}$ with $f(0) = f(n+1) = 0$. Show that the restriction of f to $\{1, \ldots, n\}$ is an eigenfunction of A_{W_n} with the same eigenvalue.

3. Deduce:

$$\sigma(A_{W_n}) = \left\{ 2 \cos \frac{\pi j}{n+1} : \; j \in \{1, \ldots, n\} \right\}.$$

4. Let $f : \{0, \ldots, 2n-1\} \to \mathbb{K}$ be an eigenfunction of $L_{C_{2n}}$ with

$$f(0) = f(1), \quad f(n) = f(n+1).$$

$$1 \underline{\hspace{3cm}} 2 \underline{\hspace{3cm}} 3 \; \text{-----------} \; \cdots \; \text{-----------} \; n$$

Fig. 1.20 The path graph W_n

Fig. 1.21 Example graphs

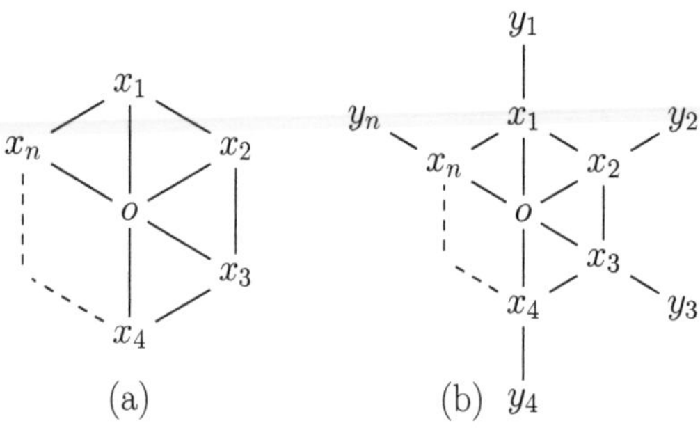

Fig. 1.22 The Petersen graph

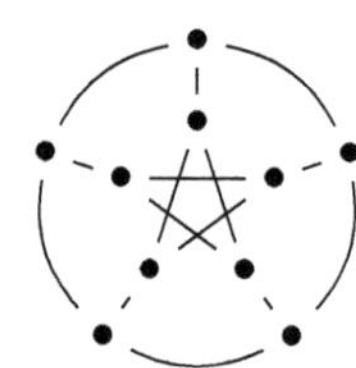

Show that the restriction of f to $\{1, \ldots, n\}$ is an eigenfunction of L_{W_n} with the same eigenvalue.

5. Deduce that

$$\sigma(L_{W_n}) = \left\{ 2 - 2\cos\frac{\pi j}{n} \; : \; j \in \{0, \ldots, n-1\} \right\}.$$

Exercise 1.6.5 Calculate the L-eigenvalues for the graphs in Fig. 1.21.

Exercise 1.6.6 Let $G = (X, E)$ and $\deg_{\max} G := \max_{x \in X} \deg x$. Show:

1. G is regular if and only if the vector $\mathbb{1} := (1, \ldots, 1)$ is an eigenvector of A.
2. A connected graph G is regular if and only if $\deg_{\max} G$ is an eigenvalue of A.
3. A k-regular graph is connected if and only if k is a simple A-eigenvalue.

Exercise 1.6.7 The Petersen graph is shown in Fig. 1.22.

1. Calculate its A-spectrum using symmetries.
2. Calculate its A-spectrum using the identity $A^2 + A = 2I + J$. (The identity must first be proven!)

Can the edges of $K_{10} =: (X, E)$ be divided into three disjoint families such that the edges of each family form the Petersen graph? (In other words: can K_{10} be covered with three Petersen graphs?)

Assume that a decomposition $E = E_1 \sqcup E_2 \sqcup E_3$ with the above properties exists, then $P_i := (X, E_i)$ are Petersen graphs. Let A_i be their adjacency matrices.

3. Show the identity $A_1 + A_2 + A_3 = J - I$.
4. Show: $\ker(A_i - I) \subset \mathbb{1}^{\perp}$ for $i \in \{1, 2, 3\}$.
5. Show: $\ker(A_1 - I) \cap \ker(A_2 - I) \neq \{0\}$.
6. Let $f \in \ker(A_1 - I) \cap \ker(A_2 - I)$ with $f \neq 0$. Show that $A_3 f = -3f$.
7. Find a contradiction.

Exercise 1.6.8 Let $k, \lambda, \mu \in \mathbb{N}_0$. A graph $G = (X, E)$ is *strongly regular* with parameters (k, λ, μ), if it satisfies the following three properties:

- G is *k-regular*.
- Any two adjacent vertices have exactly λ common neighbors.
- Any two non-adjacent vertices have exactly μ common neighbors.

1. Show: the adjacency matrix A of a strongly regular graph with parameters (k, λ, μ) satisfies

$$A^2 = kI + \lambda A + \mu(J - I - A). \qquad (1.22)$$

2. Inverse: Let G be a graph such that the identity (1.22) holds for some $k, \lambda, \mu \in \mathbb{R}$. Show: $k, \lambda, \mu \in \mathbb{N}_0$ and G is strongly regular with parameters (k, λ, μ).
3. Show: A strongly regular graph has at most three (pairwise different) A-eigenvalues.
   Hint: Use an orthonormal basis (ψ_j) of eigenvectors of A.
4. Inverse: Let G be a connected k-regular graph and let k, a, b be pairwise different, such that every A-eigenvalue is an element of $\{k, a, b\}$.
   Show: There exists a $c \in \mathbb{R}$ with $(A - aI)(A - bI) = cJ$ and deduce that G is strongly regular.

1.7 Friendship Theorem

We want to prove a mathematical version of the following statement: If at a party every two people have exactly one friend in common, then there is a person who is everyone else's friend.[3] The theorem was found in 1966 by Erdős, Rényi, Sós [49] and proven using complex combinatorial methods. The proof was later simplified: The version presented here is from [3, Chapter 44].

Theorem 1.7.1 *Let $G = (X, E)$ be a graph with at least three vertices and the following property: Any two vertices have exactly one common neighbor vertex. Then there is a vertex that is adjacent to all other vertices, and G is a so-called windmill graph (Fig. 1.23).*

---

[3] Note for the instructors: The results of this section are not used in the rest of the text.

$$X = \{o, a_1, b_1, \ldots, a_m, b_m\},$$

$$E = \bigcup_{j=1}^{m} \{\{o, a_j\}, \{o, b_j\}, \{a_j, b_j\}\}.$$

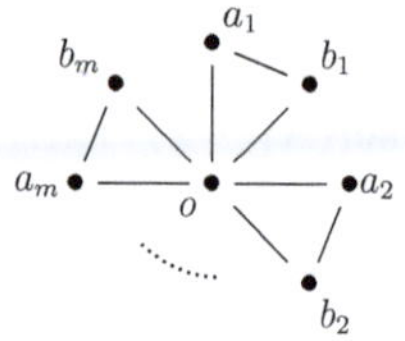

Fig. 1.23 Windmill graph

Proof Let $n := |X|$ and let A and D be the adjacency matrix and the degree matrix of G. For $n = 3$ the statement is trivial, so we assume $n \geq 4$.

The condition on G can be reformulated as follows: Between any $x, y \in X$ with $x \neq y$ there is exactly one edge path of length 2 (in particular, G is connected). This means that $(A^2)_{x,y} = 1$ for all $x \neq y$. Recall, $(A^2)_{x,x} = \deg x \equiv D_{x,x}$, so we have in total

$$A^2 = D + J - I, \quad J := J_n, \quad I := I_n$$

Since A and A^2 commute, we get $A(D + J - I) = (D + J - I)A$, i.e.

$$AD + AJ = DA + JA. \tag{1.23}$$

This equality can be written componentwise:

$$(AD)_{x,y} = \sum_{z \in X} A_{x,z} D_{z,y} = \sum_{z \in X} A_{x,z} \underbrace{\delta_{z,y} \deg y}_{D_{z,y}} = A_{x,y} \deg y,$$

$$(DA)_{x,y} = \sum_{z \in X} D_{x,z} A_{z,y} = \sum_{z \in X} \underbrace{\delta_{x,z} \deg x}_{D_{x,z}} A_{z,y} = A_{x,y} \deg x,$$

$$(AJ)_{x,y} = \sum_{z \in X} A_{x,z} J_{z,y} = \sum_{z \in X} A_{x,z} = \sum_{z:\, z \sim x} 1 = \deg x,$$

$$(JA)_{x,y} = \sum_{z \in X} J_{x,z} A_{z,y} = \sum_{z \in X} A_{z,y} = \sum_{z:\, z \sim y} 1 = \deg y.$$

With (1.23) we obtain $A_{x,y} \deg y + \deg x = A_{x,y} \deg x + \deg y$, i.e.

$$(1 - A_{x,y})(\deg x - \deg y) = 0 \quad \text{for all } x, y \in X.$$

If $x \not\sim y$, then $A_{x,y} = 0$, so $\deg x = \deg y$.

Now we conduct a proof by contradiction. Assume:

$$\text{There is no vertex that is adjacent to all other vertices.} \tag{1.24}$$

This first implies that G is not a complete graph, so there are $x, y \in X$ with $x \nsim y$, and from this follows $\deg x = \deg y$. According to the assumption, x and y have a common neighbor, which we denote by o: $o \in X$ with $o \sim x$ and $o \sim y$. Every vertex $z \in X \setminus \{x, y, o\}$ is not a common neighbor of x and y, i.e., we have $z \nsim x$ or $z \nsim y$ and from this follows $\deg x = \deg y = \deg z$. In particular, z can be chosen with $z \nsim o$ (since o is not adjacent to all vertices according to the assumption), then we get $\deg o = \deg z$. It follows that all vertices of A have the same degree, so G is regular.

Let $k \in \mathbb{N}$ with $\deg x = k$ for all $x \in X$. The case $k = 1$ can be excluded (all 1-regular graphs are disjoint unions of several copies of K_2, which obviously do not satisfy the original assumption), so $k \geq 2$.

From $D = kI$ and (1.23) it follows $A^2 = (k-1)I + J$. We multiply this equality from the right with the column $\mathbb{1}$. Due to the k-regularity of G we have $A\mathbb{1} = k\mathbb{1}$ and $A^2\mathbb{1} = k^2\mathbb{1}$, while $J\mathbb{1} = n\mathbb{1}$, and it follows

$$k^2 = k - 1 + n. \tag{1.25}$$

Let $F := \mathbb{1}^\perp \subset \mathbb{K}^n$, then F is a $(n-1)$-dimensional invariant subspace of A and $A : F \to F$ has $n - 1$ eigenvalues. For each $f \in F$ we have $Jf = 0$, so $A^2 f = (k-1)f$. It follows that every eigenvalue λ of $A|_F$ is a solution of $\lambda^2 = k - 1$, i.e. $\lambda \in \{-\sqrt{k-1}, +\sqrt{k-1}\}$.

Thus, the spectrum of A consists of the simple eigenvalue k, the r-fold eigenvalue $-\sqrt{k-1}$ and the s-fold eigenvalue $+\sqrt{k-1}$, where $r, s \in \mathbb{N}_0$ with $r + s = n - 1$. To gain additional information about r and s, we use the trace theorem for A (Theorem 1.5.14):

$$0 = \operatorname{tr} A = k - r\sqrt{k-1} + s\sqrt{k-1} = k + (s-r)\sqrt{k-1},$$

in particular $s \neq r$ and

$$\sqrt{k-1} = \frac{k}{r-s}. \tag{1.26}$$

This implies that $\sqrt{k-1}$ is a rational number. The root of a natural number is either an integer or an irrational number, therefore,

$$h := \sqrt{k-1} \in \mathbb{N}.$$

Then $k = h^2 + 1$, we can rewrite (1.26) as

$$(r-s)h = h^2 + 1$$

and it follows that $h^2 + 1$ is divisible by h. Since h^2 is also divisible by h, the difference $(h^2 + 1) - h^2 \equiv 1$ must also be divisible by h, so $h = 1$. Then $k = 2$

and from (1.25) it follows $n = 3$. The only 2-regular graph with 3 vertices is the complete graph K_3, for which the assumption (1.24) is not fulfilled.

This contradiction shows that the statement (1.24) is false, so there is a vertex $o \in X$ that is connected to all other vertices. This proves the first part of the statement.

For each $x \in X \setminus \{o\}$ there is exactly one vertex $\varphi(x) \in X \setminus \{o, x\}$ that is connected to o and x, where $\varphi\big(\varphi(x)\big) = x$. So we can divide $X \setminus \{o\}$ into m disjoint pairs (a_j, b_j) with $b_j = \varphi(a_j)$, and by construction we have

$$\{o, a_j\}, \{o, b_j\}, \{a_j, b_j\} \in E.$$

There are no edges of the form $\{a_j, a_k\}, \{a_j, b_k\}, \{b_j, b_k\}$ with $j \neq k$: For example, from $a_j \sim a_k$ it would follow that o and a_j have two common neighbors b_j and a_k (since $o \sim a_k$), which is not allowed. Thus, we have shown that G is a windmill graph. $\qquad\square$

1.8　　Matrix-Tree Theorem

We want to prove the following famous theorem,[4] called the Matrix-Tree Theorem or Kirchhoff's Theorem, which served as one of the first motivations for spectral graph theory:

Theorem 1.8.1 *Let*

- $G := (X, E)$ *be a graph with $n \geq 2$ vertices,*
- $N(G)$ *be the number of spanning trees in G,*
- $0 = \lambda_1 \leq \lambda_2 \leq \cdots \leq \lambda_n$ *be the eigenvalues of the Laplacian matrix L for G.*

Then

$$N(G) = \frac{\lambda_2 \cdot \ldots \cdot \lambda_n}{n}. \tag{1.27}$$

The number $N(G)$ of spanning trees in a graph G is another example of a graph invariant, and we want to show that this invariant can be read from the eigenvalues of the associated Laplace matrix. For disconnected G, the formula (1.27) is trivially true: G has no spanning trees, so $N(G) = 0$, and also $\lambda_2 = \lambda_1 = 0$ (since G has at least two connected components) and thus the product on the right-hand side of (1.27) is also zero.

The proof for the general case is broken down into several small steps. In particular, we will derive further formulas for $N(G)$ during the proof (see Corollaries 1.8.7 and 1.8.9 below).

---

[4] Note for instructors: The results of this section are not used in the rest of the text.

First, some new objects are introduced. We choose an orientation on G and denote by $\vec{V}$ the corresponding oriented edge-vertex matrix. For $a \in X$, let

$$X_a := X \setminus \{a\}, \qquad \vec{V}[a] := \text{the matrix } \vec{V} \text{ without the } a\text{-column},$$

i.e., $\vec{V}[a]$ is an $E \times X_a$ matrix with

$$\vec{V}[a]_{e,x} = \vec{V}_{e,x} \text{ for all } e \in E,\ x \in X_a.$$

For $S \subset E$, we denote by $\vec{V}[a]_S$ the $S \times X_a$ matrix $\vec{V}[a]$ in which only the e-rows with $e \in S$ are kept:

$$(\vec{V}[a]_S)_{e,x} = \vec{V}[a]_{e,x} = \vec{V}_{e,x} \text{ for all } e \in S,\ x \in X_a.$$

Lemma 1.8.2 *Let $a \in X$ and $S \subset E$ with $|S| = n - 1$. Then:*

$$\left| \det \vec{V}[a]_S \right| = \begin{cases} 1, & \text{if } (X, S) \text{ is a spanning tree of } G, \\ 0, & \text{otherwise.} \end{cases} \tag{1.28}$$

Note that the result is independent of the choice of a!

Proof The statement is independent of the chosen orientation: When changing the orientation, some rows of $\det \vec{V}[a]_S$ are multiplied by -1, but the absolute value of the determinant remains the same.

Assume that (X, S) is not a tree. Then (Proposition 1.1.19) the graph (X, S) contains a cycle $(x_0, \ldots, \ldots, x_m = x_0)$ with $x_k \in X$, i.e., the traversed edges $e_k := \{x_{k-1}, x_k\}$ with $k \in \{1, \ldots, m\}$ are in S. We assume that $\iota e_k = x_{k-1}$ for all k (thus we have adjusted the orientation).

Consider now

$$s_x := \sum_{k=1}^{n} \vec{V}[a]_{e_k, x}, \qquad x \in X_a.$$

For $x \notin \{x_0, \ldots, x_m\}$ one has $x \not\sim e_k$ and $\nabla[a]_{e_k, x} = 0$ for all k, so $s_x = 0$. For $x = x_j$, $\vec{V}[a]_{e_j, x} = -1$ and $\vec{V}[a]_{e_{j+1}, x} = +1$, where all other terms in the last sum are zero, so again $s_x = 0$. So $s_x = 0$ for all $x \in X_a$: This means that the sum of the e_k-rows of $\vec{V}[a]$ is a zero row. Therefore, the rows of $\vec{V}[a]$ are linearly dependent and its determinant is zero.

Now let (X, S) be a spanning tree of G, then G is connected. We perform induction on n (the number of vertices in G).

For $n = 2$, $G = K_2$, then $\vec{V} = (\pm 1 \ \mp 1)$. The only possibility for $S \subset E$ with $|S| = 1$ is $S = E$ and $(X, S) \equiv G$ is a spanning tree of G. For each $a \in X$, then

$$\vec{V}[a]_S = \vec{V}[a] = (\pm 1), \quad \left| \det \vec{V}[a]_S \right| = |\pm 1| = 1.$$

Let now $n \in \mathbb{N}$ with $n \geq 3$, such that (1.28) holds for all graphs with at most $n - 1$ vertices. Since $H := (X, S)$ is a tree with $n \geq 3$ vertices, there are at least two pendant vertices in H (see Exercise 1.1.1). In particular, there is a vertex $x \in X$ with $x \neq a$ (thus $x \in X_a$), which has a single neighbor $y \in X$ in H. The edge $e := \{x, y\}$ is, according to the construction, in S and the x-column of $\vec{V}[a]_S$ looks as follows: In the e-row there is ± 1 (depending on the orientation) and all other coefficients are zero. Thus, we can expand $\det \vec{V}[a]_S$ along to the x-column:

$$\det \vec{V}[a]_S = \pm \det \vec{V}[a, x]_{S\setminus\{e\}}, \tag{1.29}$$

where the matrix $\vec{V}[a, x]_{S\setminus\{e\}}$ is obtained by deleting the e-row and the x-column in $\vec{V}[a]_S$.

Thanks to the special choice of x,

$$H' := H \setminus x \equiv \left(X \setminus \{x\}, S \setminus \{e\} \right)$$

is a spanning tree in $G' := G \setminus x$. Let $\vec{V}'$ be the oriented edge-vertex matrix for G' (where we keep the same orientation as in G), then

$$\vec{V}[a, x]_{S\setminus\{e\}} = \vec{V}'[a]_{S\setminus\{e\}}.$$

Since G' is a graph with $n - 1$ vertices and $|S \setminus \{e\}| = n - 2$, we obtain from the induction assumption $\left| \det \vec{V}'[a]_{S\setminus\{e\}} \right| = 1$. From (1.29) it follows $\left| \det \vec{V}[a]_S \right| = 1$.

This proves the statement for all n. $\qquad\square$

Corollary 1.8.3 *For every $a \in X$ it holds*

$$N(G) = \sum_{S \subset E,\, |S|=n-1} \left(\det \vec{V}[a]_S \right)^2. \tag{1.30}$$

Proof For each subset $S \subset E$ with $|S| = n - 1$ set

$$k_S := \begin{cases} 1, & \text{if } (X, S) \text{ is a spanning tree of } G, \\ 0, & \text{otherwise.} \end{cases}$$

Since every spanning tree of X is uniquely determined by the $(n - 1)$-element set of its edges, it holds

$$N(G) = \sum_{S \subset E,\, |S|=n-1} k_S,$$

and $k_S = \left(\det \vec{V}[a]_S \right)^2$ thanks to Lemma 1.8.2. $\qquad\square$

Now the goal is to write the right-hand side of (1.30) in a different form. For this, we will apply a formula from matrix analysis.

Lemma 1.8.4 (Binet-Cauchy Formula)   *Let B be a real $m \times n$ matrix, where $m \geq n$. For a non-empty subset $S \subset \{1, \ldots, m\}$ denote*

$$B_S := \text{the matrix } B \text{ in which only the rows with indices in } S \text{ are kept.}$$

Then

$$\det(B^* B) = \sum_{S \subset \{1,\ldots,m\},\ |S|=n} \left(\det B_S \right)^2.$$

Proof   We first use the explicit formula for the determinant of the $n \times n$ matrix $B^* B$:

$$\det(B^* B) = \sum_{\sigma \in S_n} \text{sgn}\,\sigma \prod_{i=1}^{n} (B^* B)_{i\sigma(i)} = \sum_{\sigma \in S_n} \text{sgn}\,\sigma \prod_{i=1}^{n} \sum_{k=1}^{m} \underbrace{B_{ki}}_{\equiv (B^*)_{ik}} B_{k\sigma(i)}$$

$$= \sum_{\sigma \in S_n} (\text{sgn}\,\sigma) \left(\sum_{k_1=1}^{m} B_{k_1 1} B_{k_1 \sigma(1)} \right) \cdot \ldots \cdot \left(\sum_{k_n=1}^{m} B_{k_n n} B_{k_n \sigma(n)} \right)$$

$$= \sum_{k_1=1}^{m} \cdots \sum_{k_n=1}^{m} B_{k_1 1} B_{k_2 2} \cdot \ldots \cdot B_{k_n n} \sum_{\sigma \in S_n} (\text{sgn}\,\sigma)$$

$$B_{k_1 \sigma(1)} B_{k_2 \sigma(2)} \cdot \ldots \cdot B_{k_n \sigma(n)}.$$

Note that

$$\sum_{\sigma \in S_n} (\text{sgn}\,\sigma) B_{k_1 \sigma(1)} B_{k_2 \sigma(2)} \cdot \ldots \cdot B_{k_n \sigma(n)} = \det B_{k_1,\ldots,k_n},$$

where $B_{k_1,\ldots,k_n}$ is the $n \times n$ matrix whose j-th row coincides with the k_j-th row of B. If $k_1, \ldots, k_n$ are not all different, $B_{k_1,\ldots,k_n}$ has two identical rows and thus $\det B_{k_1,\ldots,k_n} = 0$. If $k_1, \ldots, k_n$ are pairwise different, the unique representation

$$(k_1, \ldots, k_n) = (\ell_{\tau(1)}, \ldots, \ell_{\tau(n)}), \quad 1 \leq \ell_1 < \cdots < \ell_n \leq m, \quad \tau \in S_n,$$

holds, and we obtain

$$\det(B^* B) = \sum_{k_1=1}^{m} \cdots \sum_{k_n=1}^{m} B_{k_1 1} B_{k_2 2} \cdot \ldots \cdot B_{k_n n} \det B_{k_1,\ldots,k_n}$$

$$= \sum_{\substack{k_1,\ldots,k_n \in \{1,\ldots,m\} \\ \text{pairwise different}}} B_{k_1 1} B_{k_2 2} \cdot \ldots \cdot B_{k_n n} \det B_{k_1,\ldots,k_n}$$

$$
\begin{aligned}
&= \sum_{1 \le \ell_1 < \cdots < \ell_n \le m} \ \sum_{\tau \in S_n} B_{\ell_{\tau(1)}1} \cdot \ldots \cdot B_{\ell_{\tau(n)}n} \det \underbrace{B_{\ell_{\tau(1)},\ldots,\ell_{\tau(n)}}}_{=(\operatorname{sgn}\tau)\,\det B_{\{\ell_1,\ldots,\ell_n\}}}
\end{aligned}
$$

$$
= \sum_{1 \le \ell_1 < \cdots < \ell_n \le m} \left(\sum_{\tau \in S_n} (\operatorname{sgn}\tau) B_{\ell_{\tau(1)}1} \cdot \ldots \cdot B_{\ell_{\tau(n)}n} \right) \det B_{\{\ell_1,\ldots,\ell_n\}}
$$

$$
= \sum_{1 \le \ell_1 < \cdots < \ell_n \le m} \det\!\left(B_{\{\ell_1,\ldots,\ell_n\}}{}^{*}\right) \det B_{\{\ell_1,\ldots,\ell_n\}}
$$

$$
= \sum_{1 \le \ell_1 < \cdots < \ell_n \le m} \left(\det B_{\{\ell_1,\ldots,\ell_n\}} \right)^2,
$$

which is exactly the formula we were looking for. $\qquad\square$

Remark 1.8.5 (Geometric Interpretation of the Binet-Cauchy Formula) If B is
a real $m \times n$ matrix and $b^1, \ldots, b^n \in \mathbb{R}^m$ are the columns of B, then

$$
\det(B^* B) = (\operatorname{vol}_n P)^2,
$$

where P is the n-dimensional parallelepiped spanned by $b^1, \ldots, b^n$ (this formula
plays a central role in analysis lectures when integrating over submanifolds).

For $S \subset \{1, \ldots, m\}$ with $|S| = n$ and each $x \in \mathbb{R}^m$ we get $x_S \in \mathbb{R}^n$ by keeping
only the coordinates in x with indices in S. Then $b_S^1, \ldots, b_S^n$ are exactly the columns
of B_S and $|\det B_S|$ is the n-dimensional volume of the parallelepiped P_S spanned
by $b_S^1, \ldots, b_S^n$. The vectors b_S^j can be identified with the orthogonal projections of
the b^j onto the n-dimensional coordinate hyperplane

$$
X_S := \{(x_1, \ldots, x_m) \in \mathbb{R}^m : x_j = 0 \text{ for all } j \notin S\}
$$

and P_S can then be seen as the orthogonal projection of P. The Binet-Cauchy
formula then takes the form

$$
(\operatorname{vol}_n P)^2 = \sum_S (\operatorname{vol}_n P_S)^2,
$$

where we sum over all n-element S and thus over all n-dimensional coordinate
hyperplanes. In other words: The square of the n-dimensional volume of an n-
dimensional parallelepiped in $\mathbb{R}^m$ is the sum of the squares of the n-dimensional
volumes of the orthogonal projections of the parallelepiped onto the n-dimensional
coordinate hyperplanes ("Pythagorean theorem").

For $a, b \in X$ denote

$$
L[a, b] := \text{the matrix } L \text{ without the } a\text{-row and without the } b\text{-column.}
$$

In other words, $L[a, b]$ is the $X_a \times X_b$ matrix with

$$L[a, b]_{x,y} = L_{x,y} \text{ for all } x \in X_a \text{ and } y \in X_b.$$

Lemma 1.8.6 *For all $a \in X$ one has*

$$L[a, a] = \vec{\nabla}[a]^* \vec{\nabla}[a]. \tag{1.31}$$

Proof For $x, y \in X_a$ it holds that

$$
\left(\vec{\nabla}[a]^* \vec{\nabla}[a] \right)_{x,y} = \sum_{e \in E} \left(\vec{\nabla}[a]^* \right)_{x,e} \vec{\nabla}[a]_{e,y} = \sum_{e \in E} \vec{\nabla}[a]_{e,x} \vec{\nabla}[a]_{e,y}
$$

$$
= \sum_{e \in E} \vec{\nabla}_{e,x} \vec{\nabla}_{e,y} = \sum_{e \in E} (\vec{\nabla}^*)_{x,e} \vec{\nabla}_{e,y} = \left(\vec{\nabla}^* \vec{\nabla} \right)_{x,y}
$$

$$
= L_{x,y} = L[a, a]_{x,y}.
$$

$\square$

Corollary 1.8.7 *For each $a \in X$ one has $N(G) = \det L[a, a]$.*

Proof When applying the Binet-Cauchy formula (Lemma 1.8.4) to the representation (1.31), we get

$$
\det L[a, a] = \sum_{S \subset E, \, |S| = n-1} \left(\det \vec{\nabla}[a]_S \right)^2,
$$

which according to Corollary 1.8.3 coincides with $N(G)$. $\square$

Now we want to relate $L[a, a]$ to the eigenvalues of L.

Lemma 1.8.8 *For each $a \in X$ it holds*

$$
\det L[a, a] = \frac{\lambda_2 \cdot \ldots \cdot \lambda_n}{n}.
$$

Proof Thanks to Corollary 1.8.7, the number $N := \det L[a, a]$ is independent of the choice of a. We have

$$
nN = \sum_{a \in X} \det L[a, a] = \text{the sum of the principal minors of } L \text{ of order } n - 1
$$

$$
= \sum_{1 \leq k_1 < \cdots < k_{n-1} \leq n} \lambda_{k_1} \cdot \ldots \cdot \lambda_{k_{n-1}}.
$$

Since $\lambda_1 = 0$, all summands with $k_1 = 1$ are zero, so

$$nN = \sum_{2 \leq k_1 < \cdots < k_{n-1} \leq n} \lambda_{k_1} \cdot \ldots \cdot \lambda_{k_{n-1}}.$$

This sum contains a single summand with $(k_1, \ldots, k_n) = (2, \ldots, n)$, so

$$nN = \lambda_2 \cdot \ldots \cdot \lambda_n.$$

$\square$

Proof of Theorem 1.8.1 Combine Corollary 1.8.7 with Lemma 1.8.8. $\square$

One can also find further representations for $N(G)$.

Corollary 1.8.9 *It holds that* $N(G) = \big| \det L[a, b] \big|$ *for all* $a, b \in X$.

Proof Replace the a-row of L with the row

$$0, \ldots, 0, \ \underbrace{1}_{a\text{-column}}, 0, \ldots, 0, \ \underbrace{-1}_{b\text{-column}}, 0, \ldots, 0,$$

thereby creating a new matrix L'. The sum of the coefficients in each row of L' is zero, so $L'\mathbb{1} = 0$ and L' is not invertible, i.e. $\det L' = 0$. By expanding $\det L'$ along the a-row we obtain (for a suitable $j \in \mathbb{N}$)

$$0 = \det L' = \det L[a, a] + (-1)^j \det L[a, b],$$

so $\det L[a, b] = \pm \det L[a, a]$. It follows

$$\big| \det L[a, b] \big| = \big| \det L[a, a] \big| = \big| N(G) \big| = N(G).$$

$\square$

Example 1.8.10 From the calculations in Sect. 1.6 it follows:

$$N(K_n) = n^{n-2}, \qquad\qquad N(K_{m,n}) = n^{m-1} m^{n-1}.$$

Remark 1.8.11 If an edge $e \in E$ of a connected graph $G = (X, E)$ is contained in many spanning trees, it can be considered as "important". Therefore, the "importance" of e can be measured with the value

$$w(e) := \begin{cases} \dfrac{N(G)}{N(G \setminus e)}, & G \setminus e \text{ is connected,} \\[2ex] \infty, & G \setminus e \text{ is not connected,} \end{cases}$$

Graph Properties and Min-Max Principle

2

In this chapter, we will learn additional techniques of linear algebra, especially the min-max principle for the calculation of eigenvalues. With this, various estimates for the most important graph invariants are derived using the eigenvalues of the graph matrices.

2.1 Courant-Fischer Theorem

Let M be a self-adjoint linear mapping in a vector space V and let $v \in V$ with $v \neq 0$. The number

$$\frac{\langle v, Mv \rangle}{\langle v, v \rangle}$$

is often referred to as the **Rayleigh quotient** of v. If v is an eigenvector of M with eigenvalue λ, then

$$\frac{\langle v, Mv \rangle}{\langle v, v \rangle} = \frac{\langle v, \lambda v \rangle}{\langle v, v \rangle} = \frac{\lambda \langle v, v \rangle}{\langle v, v \rangle} = \lambda.$$

The central idea of the Courant-Fischer theorem is that the eigenvalues of M can be estimated by minimizing and maximizing the Rayleigh quotients over certain subspaces of V. We first show that the Rayleigh quotient has a maximum and a minimum on each subspace: This simple property is used in many places.

Lemma 2.1.1 *Let* $U \subset V$ *be a subspace with* $U \neq \{0\}$. *Then there exist* $u_{\min}, u_{\max} \in U \setminus \{0\}$ *such that*

$$\frac{\langle u_{\min}, Mu_{\min} \rangle}{\langle u_{\min}, u_{\min} \rangle} \leq \frac{\langle u, Mu \rangle}{\langle u, u \rangle} \leq \frac{\langle u_{\max}, Mu_{\max} \rangle}{\langle u_{\max}, u_{\max} \rangle} \text{ for all } u \in U \setminus \{0\}.$$

© The Author(s), under exclusive license to Springer Nature Switzerland AG 2025
K. Naderi, K. Pankrashkin, *Introduction to Spectral Graph Theory*, Compact
Textbooks in Mathematics, https://doi.org/10.1007/978-3-032-01708-6_2

Proof Let $(u_1, \ldots, u_m)$ be an orthonormal basis in U. Consider the function

$$f : \mathbb{K}^m \ni (x_1, \ldots, x_m) \mapsto \langle x_1 u_1 + \ldots + x_m u_m, M(x_1 u_1 + \ldots + x_m u_m) \rangle$$

$$\equiv \sum_{j,k=1}^{m} \langle u_j, M u_k \rangle \overline{x_j} x_k \in \mathbb{R},$$

which is obviously continuous. Since the unit sphere

$$\mathbb{S} := \left\{ (x_1, \ldots, x_m) \in \mathbb{K}^m : |x_1|^2 + \cdots + |x_m|^2 = 1 \right\}$$

is compact, there exist $a = (a_1, \ldots, a_m) \in \mathbb{K}^m$ and $b = (b_1, \ldots, b_m) \in \mathbb{K}^m$ such that

$$f(a) \leq f(x) \leq f(b) \text{ for all } x \in \mathbb{S} \tag{2.1}$$

(Weierstrass theorem of minimum and maximum). For $u_{\min} := a_1 u_1 + \cdots + a_m u_m$ and $u_{\max} := b_1 u_1 + \cdots + b_m u_m$, we have $u_{\min}, u_{\max} \in U$ with

$$\langle u_{\min}, u_{\min} \rangle \equiv \|u_{\min}\|^2 = |a_1|^2 + \cdots + |a_m|^2 = 1,$$

$$\langle u_{\max}, u_{\max} \rangle \equiv \|u_{\max}\|^2 = |b_1|^2 + \cdots + |b_m|^2 = 1,$$

$$f(a) = \frac{\langle u_{\min}, M u_{\min} \rangle}{\langle u_{\min}, u_{\min} \rangle}, \quad f(b) = \frac{\langle u_{\max}, M u_{\max} \rangle}{\langle u_{\max}, u_{\max} \rangle}.$$

Since the linear combinations $x_1 u_1 + \ldots + x_m u_m$ with $(x_1, \ldots, x_m) \in \mathbb{S}$ cover the unit sphere $\mathbb{S}_U := \{u \in U : \|u\| = 1\}$ in U, it follows from (2.1)

$$f(a) \leq \langle u, M u \rangle \leq f(b) \text{ for all } u \in \mathbb{S}_U.$$

Let $u \in U \setminus \{0\}$, then $u / \|u\| \in \mathbb{S}_U$, so

$$\frac{\langle u_{\min}, M u_{\min} \rangle}{\langle u_{\min}, u_{\min} \rangle} \equiv f(a) \leq \underbrace{\left\langle \frac{u}{\|u\|}, M \frac{u}{\|u\|} \right\rangle}_{= \frac{\langle u, Mu \rangle}{\langle u, u \rangle}} \leq f(b) \equiv \frac{\langle u_{\max}, M u_{\max} \rangle}{\langle u_{\max}, u_{\max} \rangle}.$$

$$\square$$

Theorem 2.1.2 (Courant-Fischer, also Min-Max or Max-Min Principle)   *Let*

- *V be an n-dimensional vector space with scalar product,*
- *$M : V \to V$ be a self-adjoint linear mapping,*
- *$\lambda_1(M) \leq \lambda_2(M) \leq \cdots \leq \lambda_n(M)$ be the eigenvalues of M, counted with multiplicities.*

Then for each $k \in \{1, \ldots, n\}$ one has:

- *the **min-max formula***

$$\lambda_k(M) = \min_{\substack{U \subset V \text{ subspace} \\ \dim U = k}} \max_{\substack{u \in U \\ u \neq 0}} \frac{\langle u, Mu \rangle}{\langle u, u \rangle}, \tag{2.2}$$

- *the **max-min formula***

$$\lambda_k(M) = \max_{u_1, \ldots, u_{k-1} \in V} \min_{\substack{u \in \{u_1, \ldots, u_{k-1}\}^\perp \\ u \neq 0}} \frac{\langle u, Mu \rangle}{\langle u, u \rangle}. \tag{2.3}$$

Proof We write $\lambda_j := \lambda_j(M)$. According to the spectral theorem (Theorem 1.4.12), there is an orthonormal basis $v_1, \ldots, v_n$ of V with $Mv_j = \lambda_j v_j$ for all j. Every $v \in V$ is written in the form

$$v := \sum_{j=1}^{n} z_j v_j, \quad (z_1, \ldots, z_n) \in \mathbb{K}^n,$$

then we have

$$Mv = M\Big(\sum_{j=1}^{n} z_j v_j\Big) = \sum_{j=1}^{n} z_j Mv_j = \sum_{j=1}^{n} \lambda_j z_j v_j,$$

$$\langle v, v \rangle = \sum_{j=1}^{n} |z_j|^2, \qquad \langle v, Mv \rangle = \sum_{j=1}^{n} \lambda_j |z_j|^2. \tag{2.4}$$

Let $k \in \{1, \ldots, n\}$ and let $U \subset V$ be a k-dimensional subspace. For the subspace

$$W := \operatorname{span}\{v_k, v_{k+1}, \ldots, v_n\} \equiv \{v \in V : z_1 = z_2 = \cdots = z_{k-1} = 0\}$$

we have $\dim W = n - (k - 1) = n - k + 1$, so

$$\dim U + \dim W = k + n - k + 1 = n + 1 > n = \dim V.$$

It follows that $\dim(U \cap W) \geq 1$, so there is a vector $v \in U \cap W$ with $v \neq 0$,

$$v = \sum_{j=k}^{n} z_j v_j, \quad (z_k, \ldots, z_n) \neq (0, \ldots, 0).$$

Using (2.4) we obtain

$$\max_{\substack{u\in U\\u\neq 0}}\frac{\langle u,Mu\rangle}{\langle u,u\rangle}\geq\frac{\langle v,Mv\rangle}{\langle v,v\rangle}=\frac{\sum\limits_{j=k}^{n}\lambda_j|z_j|^2}{\sum\limits_{j=k}^{n}|z_j|^2}\geq\frac{\sum\limits_{j=k}^{n}\lambda_k|z_j|^2}{\sum\limits_{j=k}^{n}|z_j|^2}=\lambda_k;$$

note that the maximum on the left-hand side exists according to Lemma 2.1.1. Since this estimate holds for every k-dimensional subspace U, we have proven the inequality

$$\lambda_k\leq\inf_{\substack{U\subset V\ \text{subspace}\\\dim U=k}}\ \max_{\substack{u\in U\\u\neq 0}}\frac{\langle u,Mu\rangle}{\langle u,u\rangle}\tag{2.5}$$

Now consider the k-dimensional subspace $U':=\mathrm{span}\{v_1,\ldots,v_k\}$. For all $u\in U'$ with $u\neq 0$ we have

$$u=\sum_{j=1}^{k}z_jv_j,\quad z_1,\ldots,z_k\in\mathbb{K},\quad(z_1,\ldots,z_k)\neq(0,\ldots,0),$$

$$\langle u,u\rangle=\sum_{j=1}^{k}|z_j|^2,\quad\langle u,Mu\rangle=\sum_{j=1}^{k}\lambda_j|z_j|^2\leq\sum_{j=1}^{k}\lambda_k|z_j|^2,$$

$$\frac{\langle u,Mu\rangle}{\langle u,u\rangle}\leq\frac{\lambda_k\sum\limits_{j=1}^{k}|z_j|^2}{\sum\limits_{j=1}^{k}|z_j|^2}=\lambda_k.$$

It follows

$$\inf_{\substack{U\subset V\ \text{subspace}\\\dim U=k}}\ \max_{\substack{u\in U\\u\neq 0}}\frac{\langle u,Mu\rangle}{\langle u,u\rangle}\leq\max_{\substack{u\in U'\\u\neq 0}}\frac{\langle u,Mu\rangle}{\langle u,u\rangle}\leq\lambda_k\overset{(2.5)}{\leq}\inf_{\substack{U\subset V\ \text{subspace}\\\dim U=k}}\ \max_{\substack{u\in U\\u\neq 0}}\frac{\langle u,Mu\rangle}{\langle u,u\rangle},$$

and thus all numbers in this chain of inequalities are equal,

$$\lambda_k=\max_{\substack{u\in U'\\u\neq 0}}\frac{\langle u,Mu\rangle}{\langle u,u\rangle}=\inf_{\substack{U\subset V\ \text{subspace}\\\dim U=k}}\ \max_{\substack{u\in U\\u\neq 0}}\frac{\langle u,Mu\rangle}{\langle u,u\rangle},$$

in particular, one can replace inf by min. The min-max formula (2.2) is proven.

To prove the max-min formula (2.3), we first use

$$\lambda_k(M) = \min_{j \geq k} \lambda_j(M) = \min_{\substack{U \subset V \text{ subspace} \\ \dim U \geq k}} \max_{\substack{u \in U \\ u \neq 0}} \frac{\langle u, Mu \rangle}{\langle u, u \rangle}$$

and note that we can also substitute $(-M)$ for M:

$$\lambda_k(M) = -\lambda_{n+1-k}(-M) = - \min_{\substack{U \subset V \text{ subspace} \\ \dim U \geq n+1-k}} \max_{\substack{u \in U \\ u \neq 0}} \left(- \frac{\langle u, Mu \rangle}{\langle u, u \rangle} \right)$$

$$= \max_{\substack{U \subset V \text{ subspace} \\ \dim U \geq n+1-k}} \min_{\substack{u \in U \\ u \neq 0}} \frac{\langle u, Mu \rangle}{\langle u, u \rangle}.$$

$$(2.6)$$

A subspace $U \subset V$ satisfies $\dim U \geq n+1-k$ if and only if $\dim U^\perp \leq k-1$, i.e., if and only if $U = \{u_1, \ldots, u_{k-1}\}^\perp$ with $u_1, \ldots, u_{k-1} \in V$. If this representation is inserted into (2.6), the desired formula (2.3) is obtained. $\qquad\square$

For a self-adjoint linear mapping M, we will often write:

$$\lambda_{\max}(M) := \text{the largest eigenvalue of } M,$$

$$\lambda_{\min}(M) := \text{the smallest eigenvalue of } M,$$

which are particularly important.

Corollary 2.1.3 (Characterization of the Largest/Smallest Eigenvalue) *Let M be a self-adjoint linear mapping in a finite-dimensional vector space V with scalar product, then*

$$\lambda_{\min}(M) = \min_{v \in V,\ v \neq 0} \frac{\langle v, Mv \rangle}{\langle v, v \rangle}, \tag{2.7}$$

$$\lambda_{\max}(M) = \max_{v \in V,\ v \neq 0} \frac{\langle v, Mv \rangle}{\langle v, v \rangle}. \tag{2.8}$$

Furthermore,

$$\ker\left(M - \lambda_{\min}(M)I \right) = \left\{ v \in V : \langle v, Mv \rangle = \lambda_{\min}(M)\|v\|^2 \right\}, \tag{2.9}$$

$$\ker\left(M - \lambda_{\max}(M)I \right) = \left\{ v \in V : \langle v, Mv \rangle = \lambda_{\max}(M)\|v\|^2 \right\}. \tag{2.10}$$

Proof Let $n := \dim V$. Since $\lambda_{\min}(M) = \lambda_1(M)$, the equality (2.7) follows directly from the min-max formula (2.2) with $k = 1$. Furthermore, $\lambda_{\max}(M) = \lambda_n(M)$, so (2.8) follows from the max-min formula (2.3) with $k = n$.

Now we prove (2.9). We will abbreviate $\lambda_{\min} := \lambda_{\min}(M)$. The inclusion $\subset$ is clear: If $Mv = \lambda_{\min}v$, then

$$\langle v, Mv \rangle = \langle v, \lambda_{\min} v \rangle = \lambda_{\min} \langle v, v \rangle = \lambda_{\min} \|v\|^2.$$

Now we show the inclusion $\supset$. Let $v \in V$ with $\langle v, Mv \rangle = \lambda_{\min}(M)\|v\|^2$. Take any $u \in V$ and consider the following function $F : \mathbb{R} \to \mathbb{R}$:

$$\begin{aligned}
F(t) &:= \langle v + tu, M(v + tu) \rangle - \lambda_{\min} \|v + tu\|^2 \\
&\equiv \underbrace{\langle v, Mv \rangle - \lambda_{\min} \|v\|^2}_{=0} \\
&\quad + t\Big(\langle v, Mu \rangle + \langle u, Mv \rangle - \lambda_{\min} \langle u, v \rangle - \lambda_{\min} \langle v, u \rangle \Big) \\
&\quad + t^2 \Big(\langle u, Mu \rangle - \lambda_{\min} \|u\|^2 \Big) \\
&= 2t \operatorname{Re}\langle u, Mv - \lambda_{\min} v \rangle + t^2 \langle u, Mu - \lambda_{\min} u \rangle.
\end{aligned}$$

From the characterization (2.7) of $\lambda_{\min}$, it follows that $F(t) \geq 0$ for all $t \in \mathbb{R}$. Since $F(0) = 0$, F has a global minimum at the point $t = 0$. Since F is continuously differentiable (F is a polynomial), we have $F'(0) = 0$, i.e., $\operatorname{Re}\langle u, Mv - \lambda_{\min} v \rangle = 0$. Since $u \in V$ was arbitrary, we can set $u := Mv - \lambda_{\min} v$, then

$$0 = \operatorname{Re}\langle Mv - \lambda_{\min} v, Mv - \lambda_{\min} v \rangle = \operatorname{Re} \|Mv - \lambda_{\min} v\|^2 = \|Mv - \lambda_{\min} v\|^2,$$

so $Mv - \lambda_{\min} v = 0$ and $v \in \ker(M - \lambda_{\min} I)$. This proves (2.9), and (2.10) is proven completely analogously. $\qquad\square$

Remark 2.1.4 The significance of the Courant-Fischer theorem lies in the fact that various upper and lower estimates for the eigenvalues can be obtained by calculating the Rayleigh quotients for suitably chosen vectors v (**test vectors** or **test functions**).

Remark 2.1.5 For real symmetric matrices, one can work with subspaces in $\mathbb{C}^X$ or $\mathbb{R}^X$ in all formulas in the Courant-Fischer theorem: In both cases, the same result is obtained (see Remark 1.4.14).

We will see later that the second smallest eigenvalue can also be of interest. The following representation will be useful for this.

Corollary 2.1.6 *Let M be a self-adjoint linear mapping in a finite-dimensional vector space V with eigenvalues $\lambda_1(M) \leq \lambda_2(M) \leq \dots$. Then*

$$\lambda_2(M) = \min_{u \perp f, \, u \neq 0} \frac{\langle u, Mu \rangle}{\langle u, u \rangle},$$

where f is any eigenvector corresponding to $\lambda_1(M)$. A vector $u \in f^\perp$ is an eigenvector corresponding to $\lambda_2(M)$ if and only if the minimum is attained on u.

Proof Let f be an eigenvector corresponding to $\lambda_1(M)$ and consider $U := f^\perp$. Then $M_U : U \ni u \mapsto Mu \in U$ is a self-adjoint linear mapping in U and

$$\sigma(M) = \{\lambda_1(M)\} \cup \sigma(M_U),$$

where $\lambda_1(M)$ is the smallest element of $\sigma(M)$. From Corollary 2.1.3 it follows

$$\lambda_2(M) = \lambda_{\min}(M_U) = \min_{u \in U,\ u \neq 0} \frac{\langle u, Mu \rangle}{\langle u, u \rangle},$$

$$\ker\left(M - \lambda_2(M)I\right) = \left\{u \in U : \langle u, Mu \rangle = \lambda_2(M)\|u\|^2\right\}.$$

$\square$

With the Courant-Fischer theorem, one can compare the eigenvalues of two different linear mappings.

Corollary 2.1.7 (Monotonicity of Eigenvalues) *Let*

- *V be an n-dimensional vector space with scalar product,*
- *$M, M' : V \to V$ be self-adjoint linear mappings with*

$$\langle v, Mv \rangle \leq \langle v, M'v \rangle \text{ for all } v \in V.$$

Then $\lambda_k(M) \leq \lambda_k(M')$ for all $k \in \{1, \ldots, n\}$.

Proof Let $U \subset V$ be a k-dimensional subspace with

$$\lambda_k(M') = \max_{u \in U, u \neq 0} \frac{\langle u, M'u \rangle}{\langle u, u \rangle},$$

then

$$\lambda_k(M) \leq \max_{u \in U, u \neq 0} \frac{\langle u, Mu \rangle}{\langle u, u \rangle} \leq \max_{u \in U, u \neq 0} \frac{\langle u, M'u \rangle}{\langle u, u \rangle} = \lambda_k(M').$$

$\square$

The following statement requires a bit more work.

Corollary 2.1.8 (Interlacing Principle) *Let*

- *V be an n-dimensional vector space,*
- *$M : V \to V$ be a self-adjoint linear mapping,*

- V' be an m-dimensional vector space with $m \leq n$,
- $\Theta : V' \to V$ be a linear mapping with $\Theta^* \Theta = I$,

then

$$\lambda_k(M) \leq \lambda_k(\Theta^* M \Theta) \leq \lambda_{k+n-m}(M) \text{ for all } k \in \{1, \ldots, m\}, \tag{2.11}$$

in particular

$$\lambda_{\min}(M) \leq \lambda_{\min}(\Theta^* M \Theta) \leq \lambda_{\max}(\Theta^* M \Theta) \leq \lambda_{\max}(M). \tag{2.12}$$

Proof For each $v' \in V'$ one has

$$\|\Theta v'\|^2 = \langle \Theta v', \Theta v' \rangle = \langle v', \underbrace{\Theta^* \Theta}_{=I} v' \rangle = \langle v', v' \rangle = \|v'\|^2,$$

so Θ is injective, and for every subspace $U' \subset V'$ it holds $\dim \Theta(U') = \dim U'$. Let $k \in \{1, \ldots, n\}$, then

$$\{U : U \subset V \text{ subspace, } \dim U = k\} \supset \{\Theta(U') : U' \subset V' \text{ subspace, } \dim U' = k\},$$

and it follows

$$\lambda_k(M) = \min_{\substack{U \subset V \text{ subspace} \\ \dim U = k}} \max_{\substack{u \in U \\ u \neq 0}} \frac{\langle u, Mu \rangle}{\langle u, u \rangle} \geq \min_{\substack{U' \subset V' \text{ subspace} \\ \dim U' = k}} \max_{\substack{u \in \Theta(U') \\ u \neq 0}} \frac{\langle u, Mu \rangle}{\langle u, u \rangle}$$

$$= \min_{\substack{U' \subset V' \text{ subspace} \\ \dim U' = k}} \max_{\substack{u' \in U' \\ u' \neq 0}} \frac{\langle \Theta u', M \Theta u' \rangle}{\langle \Theta u', \Theta u' \rangle} = \min_{\substack{U' \subset V' \text{ subspace} \\ \dim U' = k}} \max_{\substack{u' \in U' \\ u' \neq 0}} \frac{\langle u', \Theta^* M \Theta u' \rangle}{\langle u', u' \rangle}$$

$$= \lambda_k(\Theta^* M \Theta).$$

This proves the first inequality in (2.11).

For the second inequality we use the identities

$$\lambda_j(\Theta^* M \Theta) = -\lambda_{m+1-j}(-\Theta^* M \Theta) \text{ for all } j \in \{1, \ldots, m\},$$

$$\lambda_j(M) = -\lambda_{n+1-j}(-M) \text{ for all } j \in \{1, \ldots, n\}.$$

Let $k \in \{1, \ldots, m\}$. Use the first identity for $j := k$ and the second one for $j := k + n - m$, then

$$\lambda_k(\Theta^* M \Theta) = -\lambda_{m+1-k}(-\Theta^* M \Theta), \quad \lambda_{k+n-m}(M) = -\lambda_{m+1-k}(-M),$$

according to the first part of the proof, $\lambda_{m+1-k}(-M) \leq \lambda_{m+1-k}(-\Theta^* M \Theta)$, which gives the sought second inequality in (2.11).

This proves (2.11). For $k \in \{1, n\}$ we obtain (2.12). □

Corollary 2.1.9 (Interlacing Principle: Matrices) *Let M be a $n \times n$ Hermitian matrix and $p \in \{1, \ldots, n-1\}$. Let $k_1, \ldots, k_p \in \{1, \ldots, n\}$ be distinct and let M' be the $(n-p) \times (n-p)$ matrix obtained from M by deleting the rows and columns with numbers in $\{k_1, \ldots, k_p\}$. Then it holds that*

$$\lambda_k(M) \le \lambda_k(M') \le \lambda_{k+p}(M) \text{ for all } k \in \{1, \ldots, n-p\},$$

in particular,

$$\lambda_{\min}(M) \le \lambda_{\min}(M') \le \lambda_{\max}(M') \le \lambda_{\max}(M).$$

Proof Let Θ be the $n \times (n-p)$ matrix obtained from the $n \times n$ identity matrix by deleting the columns with numbers in $\{k_1, \ldots, k_p\}$. The matrices M and Θ define linear mappings $M : \mathbb{C}^n \to \mathbb{C}^n$ and $\Theta : \mathbb{C}^{n-p} \to \mathbb{C}^n$, where M is self-adjoint. The adjoint mapping $\Theta^* : \mathbb{C}^n \to \mathbb{C}^{n-p}$ is given by the Hermitian transpose of Θ.

A small calculation shows that $\Theta^*\Theta = I_{n-p}$ and $\Theta^* M \Theta = M'$. For example, for $\{k_1, \ldots, k_p\} = \{1, \ldots, p\}$ the illustrative block representations are:

$$M = \begin{pmatrix} \cdot & \cdot \\ \cdot & M' \end{pmatrix}, \quad \Theta = \begin{pmatrix} 0 \\ I_{n-p} \end{pmatrix}, \quad \Theta^* = \begin{pmatrix} 0 & I_{n-p} \end{pmatrix},$$

$$\Theta^*\Theta = I_{n-p}, \quad \Theta^* M \Theta = M';$$

for the general case, one has to adjust the numbering accordingly. Therefore, we are within the scope of Corollary 2.1.8 with $m := n - p$. □

In some situations, one cannot estimate the eigenvalues directly, but their number in some locations.

Corollary 2.1.10 *Let V be a vector space with scalar product and $M : V \to V$ a self-adjoint linear mapping, then it holds that*

the number of negative eigenvalues of $M = \max \big\{ \dim U :$

$$U \subset V \text{ subspace with } \langle u, Mu \rangle < 0 \text{ for all } u \in U \setminus \{0\} \big\}.$$

Proof Denote $n :=$ the number of negative eigenvalues of M and

$$m := \max \big\{ \dim U : U \subset V \text{ subspace with } \langle u, Mu \rangle < 0 \text{ for all } u \in U \setminus \{0\} \big\}.$$

If $n = 0$, then $\lambda_1(M) \ge 0$, and then $\langle u, Mu \rangle \ge 0$ for all $u \in V$ by the min-max formula, and then $m = 0$. Let now $n \ge 1$. By the assumption $\lambda_n(M) < 0$ and by the min-max formula there is a subspace $U \subset V$ with $\dim U = n$ and

$$\lambda_n(M) = \max_{u \in U, u \neq 0} \frac{\langle u, Mu \rangle}{\|u\|^2}.$$

Then for all $u \in U \setminus \{0\}$ the inequality $\langle u, Mu \rangle \leq \lambda_n(M)\|u\|^2 < 0$ holds, and it follows $m \geq \dim U = n$.

Now we prove the reverse inequality. If $m = 0$, $\langle u, Mu \rangle \geq 0$ holds for all $u \in V$ and by the min-max formula $\lambda_1(M) \geq 0$ and thus $n = 0$. Assume now that $m \geq 1$. Let $U \subset V$ be a subspace with $\dim U = m$ and $\langle u, Mu \rangle < 0$ for all $u \in U \setminus \{0\}$. By Lemma 2.1.1 there is a $u_{\max} \in U \setminus \{0\}$ with

$$\frac{\langle u, Mu \rangle}{\langle u, u \rangle} \leq \frac{\langle u_{\max}, Mu_{\max} \rangle}{\langle u_{\max}, u_{\max} \rangle} =: c,$$

and from $u_{\max} \in U \setminus \{0\}$ it follows $c < 0$. By the min-max formula it follows $\lambda_m(M) \leq c < 0$, so M has at least m negative eigenvalues, i.e. $n \geq \dim U = m$.

$\square$

Corollary 2.1.11 (Sylvester's Law of Inertia) *Let*

- *V and V' be finite-dimensional vector spaces with scalar products,*
- *$M : V \to V$ a self-adjoint linear mapping,*
- *$\Theta : V' \to V$ a <u>surjective</u> linear mapping.*

Then M and $\Theta^* M \Theta$ have the same number of negative eigenvalues.

Proof For a linear mapping B we denote by $N_-(B)$ the number of negative eigenvalues of B.

We have $\Theta^* M \Theta(\ker \Theta) = \{0\} \subset \ker \Theta$, so $\ker \Theta$ is an invariant subspace of $\Theta^* M \Theta$. Thus, also $\widetilde{V} := (\ker \Theta)^\perp$ is an invariant subspace of $\Theta^* M \Theta$ (Lemma 1.4.11). With respect to the orthogonal decomposition $V' = (\ker \Theta) \oplus \widetilde{V}$, the mapping $\Theta^* M \Theta$ has the block representation

$$M = \begin{pmatrix} 0 & 0 \\ 0 & \widetilde{M} \end{pmatrix} \text{ for } \widetilde{M} := \left(\Theta^* M \Theta : \widetilde{V} \to \widetilde{V} \right),$$

thus $N_-(\widetilde{M}) = N_-(\Theta^* M \Theta)$ follows.

Now we use Corollary 2.1.10 and the bijectivity of $\Theta : \widetilde{V} \to V$:

$$N_-(\Theta^* M \Theta) = N_-(\widetilde{M})$$

$$= \max \left\{ \dim \widetilde{U} : \widetilde{U} \subset \widetilde{V} \text{ with } \langle \widetilde{u}, \widetilde{M}\widetilde{u} \rangle < 0 \text{ for all } \widetilde{u} \in \widetilde{U} \setminus \{0\} \right\}$$

$$= \max \left\{ \dim \widetilde{U} : \widetilde{U} \subset \widetilde{V} \text{ with } \langle \widetilde{u}, \Theta^* M \Theta \widetilde{u} \rangle < 0 \text{ for all } \widetilde{u} \in \widetilde{U} \setminus \{0\} \right\}$$

$$= \max \left\{ \dim \widetilde{U} : \widetilde{U} \subset \widetilde{V} \text{ with } \langle \Theta \widetilde{u}, M \Theta \widetilde{u} \rangle < 0 \text{ for all } \widetilde{u} \in \widetilde{U} \setminus \{0\} \right\}$$

$$(u := \Theta \widetilde{u}) \; = \max \Big\{ \dim U : U \subset V \text{ with } \langle u, Mu \rangle < 0 \text{ for all } u \in U \setminus \{0\} \Big\}$$

$$= N_-(M).$$

$\square$

Exercise 2.1.1 Let V be an n-dimensional vector space with scalar product and $M, N : V \to V$ two self-adjoint linear mappings, where $p := \dim \operatorname{ran} N \in \{1, \ldots, n-1\}$. Show:

1. $\lambda_k(M) \leq \lambda_{k+p}(M+N)$, for all $k \in \{1, \ldots, n-p\}$.
   Hint: Write $\lambda_{k+p}(M+N)$ with the max-min formula.
2. $\lambda_k(M) \geq \lambda_{k-p}(M+N)$, for all $k \in \{p+1, \ldots, n\}$.

2.2   Matrices with Parameters

In this section, we will prove some general statements about the smoothness of eigenvalues with respect to parameters. These properties are only needed in Sects. 3.5, 4.7, and 4.8.

Let

$$M(t) = \big(M_{i,j}(t)\big)_{i,j\in\{1,\ldots,n\}}$$

be Hermitian $n \times n$ matrices that depend on $t \in \Omega$, where $\Omega \subset \mathbb{R}^d$ is a non-empty open set. We investigate the dependence of the eigenvalues of $M(t)$ on the parameter t.

Theorem 2.2.1 (Continuity of Eigenvalues) *If all functions $t \mapsto M_{i,j}(t)$ are continuous in $t_0 \in \Omega$, then for each $\ell \in \{1, \ldots, n\}$ also $t \mapsto \lambda_\ell(M(t))$ is continuous in t_0.*

If $\lambda_\ell(M(t_0))$ is a simple eigenvalue of $M(t_0)$ for some ℓ, then $\lambda_\ell(M(t))$ is also a simple eigenvalue of $M(t)$ for all sufficiently small $|t - t_0|$.

Proof Let $\varepsilon > 0$. Since there are only finitely many coefficients, one can find a $\delta > 0$ such that $\big|M_{i,j}(t) - M_{i,j}(t_0)\big| < \varepsilon$ for all $t \in \Omega$ with $|t - t_0| < \delta$ and all i, j. Let $f = (f_1, \ldots, f_n) \in \mathbb{K}^n$, then for the same t

$$\Big|\langle f, M(t)f \rangle - \langle f, M(t_0)f \rangle\Big|$$

$$= \Big|\big\langle f, (M(t) - M(t_0))f \big\rangle\Big| = \Big| \sum_{i,j=1}^{n} \big(M_{i,j}(t) - M_{i,j}(t_0)\big)\overline{f_i} f_j \Big|$$

$$\leq \sum_{i,j=1}^{n} \left| M_{i,j}(t) - M_{i,j}(t_0) \right| |f_i| \, |f_j| \leq \varepsilon \sum_{i,j=1}^{n} |f_i| \, |f_j|$$

$$\leq 2\varepsilon \sum_{i,j=1}^{n} \left(|f_i|^2 + |f_j|^2 \right) = 4n\varepsilon \sum_{i=1}^{n} |f_i|^2 = 4n\varepsilon \|f\|^2,$$

thus $\langle f, M(t_0)f \rangle - 4n\varepsilon \|f\|^2 \leq \langle f, M(t)f \rangle \leq \langle f, M(t_0)f \rangle + 4n\varepsilon \|f\|^2$. From the min-max principle it follows

$$\lambda_\ell\big(M(t_0)\big) - 4n\varepsilon \leq \lambda_\ell\big(M(t)\big) \leq \lambda_\ell\big(M(t_0)\big) + 4n\varepsilon$$

for all $\ell \in \{1, \ldots, n\}$ and $t \in \Omega$ with $|t - t_0| < \delta$. Since ε is arbitrary, the first statement follows.

Assume that $\lambda_\ell\big(M(t_0)\big)$ is a simple eigenvalue for some $1 < \ell < n$, then

$$\lambda_{\ell-1}\big(M(t_0)\big) < \lambda_\ell\big(M(t_0)\big) < \lambda_{\ell+1}\big(M(t_0)\big).$$

Due to the continuity of the eigenvalues with respect to t, there exists a $\delta > 0$ such that

$$\lambda_{\ell-1}\big(M(t)\big) < \lambda_\ell\big(M(t)\big) < \lambda_{\ell+1}\big(M(t)\big)$$

for all $t \in \Omega$ with $|t - t_0| < \delta$, i.e., $\lambda_\ell\big(M(t)\big)$ is also a simple eigenvalue for these t. The cases $\ell = 1$ and $\ell = n$ are considered analogously. $\square$

Theorem 2.2.2 (Smoothness of Simple Eigenvalues) *Let all functions $M_{i,j}$ be C^k-smooth. Further, let $\ell \in \{1, \ldots, n\}$ and $t_0 \in \Omega$, such that the eigenvalue $\lambda_\ell(M(t_0))$ is simple. Then there exists an open neighborhood U of t_0 in Ω with the following properties:*

- *the function $U \ni t \mapsto \lambda_\ell\big(M(t)\big) \in \mathbb{R}$ is C^k-smooth,*
- *there exists a C^k-smooth mapping $U \ni t \mapsto f(t) \in \mathbb{K}^X$, such that $f(t)$ is an eigenvector of $M(t)$ corresponding to $\lambda_\ell\big(M(t)\big)$ for each $t \in U$.*

Proof Let $P_{M(t)}$ be the characteristic polynomial of $M(t)$. According to the assumption, we have

$$P_{M(t_0)}\big(\lambda_\ell(M(t_0))\big) = 0, \quad P'_{M(t_0)}\big(\lambda_\ell(M(t_0))\big) \neq 0.$$

The function $(t, \lambda) \mapsto P_{M(t)}(\lambda)$ is C^k-smooth and according to the implicit function theorem, there exists an open neighborhood U of t_0 and a uniquely determined C^k-smooth function $t \mapsto \lambda(t)$ with $\lambda(t_0) = \lambda_\ell(M(t_0))$, such that

$$P_{M(t)}\big(\lambda(t)\big) = 0 \text{ for all } t \in U.$$

Then $\lambda(t)$ is an eigenvalue of $M(t)$ for each $t \in U$. We can assume U to be sufficiently small such that the eigenvalue $\lambda_\ell(M(t))$ is simple for all $t \in U$ (which is possible according to Theorem 2.2.1). Then $\lambda(t) = \lambda_\ell(M(t))$ must hold for all $t \in U$, since there are no further eigenvalues of $M(t)$ in a small neighborhood of $\lambda(M(t_0))$. This implies the C^k-smoothness of $t \mapsto \lambda_\ell(M(t))$.

Now we show the existence of a C^k-smooth eigenvector. Recall that the number $P'_{M(t_0)}\big(\lambda_\ell(M(t_0))\big)$ is the coefficient p_1 in the representation

$$P_{M(t_0)}(\lambda) = \sum_{j=1}^{n} p_j \big(\lambda - \lambda_\ell(M(t_0))\big)^j,$$

and p_1 is (up to sign) the sum of the principal minors of order $n-1$ of the matrix $P_{M(t_0)} - \lambda_\ell(M(t_0))I$. Thus, at least one of the principal minors of order $n-1$ is non-zero. We can assume w.l.o.g. that the principal minor constructed on the first $n-1$ rows and columns is not zero. Let $B(t)$ be the submatrix of $P_{M(t)} - \lambda_\ell(M(t))I$ constructed on the first $n-1$ rows and columns. According to the assumption, $\det B(t_0) \neq 0$, and due to continuity, $\det B(t) \neq 0$ for all $t \in U$ (we can again assume that U is chosen sufficiently small). From now on, let $t \in U$. We have the block representation

$$M(t) - \lambda_\ell\big(M(t_0)\big)I = \begin{pmatrix} B(t) & c(t) \\ c(t)^* & \cdot \end{pmatrix},$$

where $t \mapsto c(t) \in \mathbb{C}^{n-1}$ is a C^k-smooth vector function (written as a column). Let

$$f(t) = \begin{pmatrix} f_1(t) \\ \cdots \\ f_{n-1}(t) \\ f_n(t) \end{pmatrix} \equiv \begin{pmatrix} \varphi(t) \\ f_n(t) \end{pmatrix}$$

be an eigenvector of $M(t)$ corresponding to $\lambda_\ell(M(t))$. We first want to show that $f_n(t) \neq 0$ for all $t \in U$. If $f_n(t) = 0$ for some $t \in U$, then

$$\begin{pmatrix} 0 \\ 0 \end{pmatrix} = \underbrace{\begin{pmatrix} B(t) & c(t) \\ c(t)^* & \cdot \end{pmatrix}}_{=M(t)-\lambda_\ell(M(t))I} \underbrace{\begin{pmatrix} \varphi(t) \\ 0 \end{pmatrix}}_{f(t)}, \quad \text{in particular } B(t)\varphi(t) = 0.$$

Since $B(t)$ is invertible, this would imply $\varphi(t) = 0$ and thus also $f(t) = 0$: contradiction. Therefore, $f_n(t) \neq 0$ for all $t \in U$ and we can assume without loss of generality that $f_n(t) \equiv 1$. Then

$$\begin{pmatrix} 0 \\ 0 \end{pmatrix} = \underbrace{\begin{pmatrix} B(t) & c(t) \\ c(t)^* & \cdot \end{pmatrix}}_{=M(t)-\lambda_\ell\left(M(t_0)\right)I} \underbrace{\begin{pmatrix} \varphi(t) \\ 1 \end{pmatrix}}_{f(t)},$$

In particular, $B(t)\varphi(t) + c(t) = 0$, so $\varphi(t) = -B(t)^{-1}c(t)$. Since $B(t)$ is invertible for every t and $t \mapsto B(t)$ is a C^k-smooth matrix function, $t \mapsto B(t)^{-1}$ is also a C^k-smooth matrix function. This implies the C^k-smoothness of φ and thus also of f. $\qquad\qquad\square$

Remark 2.2.3 For the sake of completeness, we will derive a formula for the (partial) derivatives $\partial_j\lambda_\ell$ under the assumptions of Theorem 2.2.2 (it is not used in this book).

Denote $\lambda(t) := \lambda_\ell(t)$. According to Theorem 2.2.2, one can construct eigenfunctions $f(t)$ for $\lambda(t)$ such that:

$$t \mapsto f(t) \text{ is smooth}, \quad \left\| f(t) \right\| = 1 \text{ for all } t.$$

From $M(t)f(t) = \lambda(t)f(t)$ it follows with the product rule

$$M'f + Mf' = \lambda'f + \lambda f', \tag{2.13}$$

with $f := f(t)$, $f' := \partial_j f(t)$ and analogously for M and λ.

We now compute the scalar product of (2.13) with f:

$$\langle f, M'f \rangle + \langle f, Mf' \rangle = \lambda'\langle f, f \rangle + \lambda\langle f, f' \rangle.$$

We have $\langle f, f \rangle = 1$ as well as $\langle f, Mf' \rangle = \langle Mf, f' \rangle = \lambda\langle f, f' \rangle$, so it follows that $\langle f, M'f \rangle = \lambda'$. The found identity

$$\partial_j\lambda(t) = \left\langle f(t), \left(\partial_j M(t)\right) f(t) \right\rangle$$

is known as the **Feynman-Hellmann formula**.

2.3   Perron-Frobenius Theorem

We need another important theorem about the largest eigenvalue of certain symmetric matrices with nonnegative coefficients. As a reminder, multiplication by symmetric matrices with real coefficients defines self-adjoint linear mappings, so one can use the characterizations of eigenvalues by Rayleigh quotients (Corollary 2.1.3).

Definition 2.3.1 Let X be a set and $f \in \mathbb{K}^X$. Then f is called:

- **nonnegative**, if $f(x) \geq 0$ for all $x \in X$ (we write $f \geq 0$),
- **positive**, if $f(x) > 0$ for all $x \in X$ (we write $f > 0$).

We denote

$$|f| : \ X \ni x \mapsto |f(x)|.$$

Theorem 2.3.2 (Perron-Frobenius, Weak Version) *Let M be a $X \times X$ symmetric matrix with <u>nonnegative coefficients</u>. Then:*

(a) $\lambda_{\max}(M) \geq |\lambda_{\min}(M)|$. In particular, $\lambda_{\max}(M)$ has the largest absolute value among all eigenvalues.

(b) If $f \in \mathbb{K}^X$ is an eigenvector of M for $\lambda_{\max}(M)$, then $|f|$ is also an eigenvector of M for $\lambda_{\max}(M)$.

Proof

(a) Let g be an eigenvector for $\lambda_{\min}$. We have $\big\||g|\big\| = \|g\| > 0$,

$$|\lambda_{\min}| \, \big\||g|\big\|^2 \big| = \big|\lambda_{\min}\|g\|^2\big| = |\langle g, Mg\rangle|$$

$$= \Big|\sum_{x,y\in X} M_{x,y}\overline{g(x)}g(y)\Big| \leq \sum_{x,y\in X} \big|M_{x,y}\overline{g(x)}g(y)\big|$$

$$\equiv \sum_{x,y\in X} M_{x,y}\big|\overline{g(x)}\big|\,|g(y)| \equiv \sum_{x,y\in X} M_{x,y}\overline{|g|(x)}\,|g|(y)$$

$$= \langle |g|, M|g|\rangle \leq \lambda_{\max}\big\||g|\big\|^2,$$

so $|\lambda_{\min}| \leq \lambda_{\max}$.

(b) We have $\big\||f|\big\| = \|f\| > 0$ and $\langle |f|, M|f|\rangle \leq \lambda_{\max}\big\||f|\big\|^2$. Furthermore,

$$\lambda_{\max}\big\||f|\big\|^2 = \lambda_{\max}\|f\|^2 = \langle f, Mf\rangle = \sum_{x,y\in X} M_{x,y}\overline{f(x)}f(y)$$

$$\leq \sum_{x,y\in X} \big|M_{x,y}\overline{f(x)}f(y)\big| \equiv \sum_{x,y\in X} M_{x,y}\big|\overline{f(x)}\big|\,|f(y)|$$

$$\equiv \sum_{x,y\in X} M_{x,y}\overline{|f|(x)}|f|(y) = \langle |f|, M|f|\rangle.$$

Thus, we have shown $\langle |f|, M|f|\rangle = \lambda_{\max}\big\||f|\big\|^2$, and it follows that $|f| \in \ker(M - \lambda_{\max}I)$ according to Corollary 2.1.3.

$\square$

Theorem 2.3.3 (Perron-Frobenius, Strong Version) *Let $G = (X, E)$ be a* <u>*connected*</u> *graph and M a real symmetric $X \times X$ matrix with the following properties:*

(i) $M_{x,y} \geq 0$ for all $x, y \in X$,
(ii) for all $x, y \in X$ with $x \neq y$ it holds:

$$M_{x,y} > 0 \text{ for } x \sim y, \quad M_{x,y} = 0 \text{ for } x \nsim y.$$

Then the following statements hold:

(a) There exists an eigenvector $f > 0$ of M for $\lambda_{\max}(M)$.
(b) It holds that $\lambda_{\max}(M) \geq 0$, and even $\lambda_{\max}(M) > 0$ for $|X| \geq 2$.
(c) $\lambda_{\max}(M)$ is a simple eigenvalue.
(d) $\lambda_{\max}(M)$ is the only eigenvalue with a nonnegative eigenvector.

We first prove a useful auxiliary statement:

Lemma 2.3.4 *Let G and M be as in Theorem 2.3.3. If $g \in \mathbb{K}^X$ is an eigenvector of M with $g \geq 0$, then even $g > 0$ holds.*

Proof Proof by contradiction. Let $g(a) = 0$ for some $a \in X$. Since g is an eigenvector, we have $g \neq 0$, so there is a vertex $b \in X$ with $g(b) \neq 0$, i.e. $g(b) > 0$ in our case. Since G is connected, there is a path $(a = x_0, x_1, \ldots, x_n = b)$ from a to b. We have $g(x_k) \geq 0$ for all k, where $g(x_0) = 0$ and $g(x_n) > 0$. Choose the <u>smallest</u> k with $g(x_k) > 0$, then $k \geq 1$ and $g(x_{k-1}) = 0$. Let λ be the eigenvalue corresponding to g, then

$$0 = \lambda g(x_{k-1}) = (Mg)(x_{k-1}) \equiv \sum_{y \in X} \underbrace{M_{x_{k-1}, y}}_{\geq 0} \underbrace{g(y)}_{\geq 0} \geq \underbrace{M_{x_{k-1}, x_k}}_{>0 \text{ by (ii)}} \underbrace{g(x_k)}_{>0} > 0,$$

which is obviously wrong. $\qquad\qquad\square$

Proof of Theorem 2.3.3 Denote $\lambda_{\max} := \lambda_{\max}(M)$ and $\lambda_{\min} := \lambda_{\min}(M)$.

(a) Let g be an eigenvector of M for $\lambda_{\max}$ and $f := |g|$. Then $f \geq 0$, and according to the weak version of the Perron-Frobenius theorem (Theorem 2.3.2), f is also an eigenvector for $\lambda_{\max}$. Thanks to Lemma 2.3.4, we have $f > 0$.
(b) The inequality $\lambda_{\max} \geq 0$ follows from Theorem 2.3.2(a). Now let $|X| \geq 2$, then M has at least one positive coefficient and

$$\lambda_{\max} \geq \frac{\langle \mathbb{1}, M\mathbb{1} \rangle}{\langle \mathbb{1}, \mathbb{1} \rangle} > 0.$$

(c) Proof by contradiction. Assume that $\dim \ker(M - \lambda_{\max} I) \geq 2$. Let $f > 0$ be an eigenvector for $\lambda_{\max}$, then there exists an eigenvector g for $\lambda_{\max}$ with $f \perp g$. Since all coefficients of f and M as well as the eigenvalue $\lambda_{\max}$ are real, it follows from $\langle f, g \rangle = 0$ and $Mg = \lambda_{\max} g$, that also the real vectors $\operatorname{Re} g$ and $\operatorname{Im} g$ lie in $\ker(M - \lambda_{\max} I)$ and are orthogonal to f. At least one of these vectors is not a zero vector (since $g = \operatorname{Re} g + i \operatorname{Im} g \neq 0$). So let $h \in \{\operatorname{Re} g, \operatorname{Im} g\}$ with $h \neq 0$, then h is a real eigenvector of M for $\lambda_{\max}$ with

$$\sum_{x \in X} \underbrace{f(x)}_{>0} h(x) \equiv \langle f, h \rangle = 0.$$

From this it follows that h must take positive and negative values: There exist $x_{\pm} \in X$ with $h(x_-) < 0$ and $h(x_+) > 0$. According to Theorem 2.3.2(b), $|h|$ is also an eigenvector for $\lambda_{\max}$ and due to

$$\left| h(x_+) \right| = h(x_+), \quad \left| h(x_-) \right| = -h(x_-), \quad h(x_{\pm}) \neq 0,$$

h and $|h|$ are linearly independent. Then the vector $h_* := |h| + h$ is also not a zero vector and it also lies in $\ker(M - \lambda_{\max} I)$. Obviously $h_* \geq 0$ (since $|a| + a \geq 0$ for all real numbers a) and with Lemma 2.3.4 we get $h_* > 0$. But this is in contradiction to

$$h_*(x_-) = \underbrace{\left| h(x_-) \right|}_{<0} + h(x_-) = -h(x_-) + h(x_-) = 0.$$

(d) Let $g \geq 0$ be an eigenvector of M with eigenvalue λ. According to Lemma 2.3.4, we then have $g > 0$. Let $f > 0$ be an eigenvector for $\lambda_{\max}$ and assume that $\lambda \neq \lambda_{\max}$, then $f \perp g$. But this is impossible:

$$\langle f, g \rangle = \sum_{x \in X} \underbrace{f(x)}_{>0} \underbrace{g(x)}_{>0} > 0.$$

So $\lambda = \lambda_{\max}$.

$$\square$$

Remark 2.3.5 Some statements of the Perron-Frobenius theorem also hold for a large class of nonsymmetric matrices with nonnegative coefficients, but the proofs are much more complex, as there is no direct analogue to the Courant-Fischer theorem for nonsymmetric matrices.

2.4   Edge and Vertex Monotonicity

Let $G = (X, E)$ be a graph on which we introduce an arbitrary orientation, and let A be its adjacency matrix. Since the Rayleigh quotient is actively used in the Min-Max principle, we write it out explicitly. For $f \in \mathbb{K}^X$ it holds

$$\|f\|^2 = \sum_{x \in X} |f(x)|^2,$$

and for the adjacency matrix A it holds

$$\langle f, Af \rangle = \sum_{x \in X} \overline{f(x)} \Big(\sum_{y:\, y \sim x} f(y) \Big) = \sum_{e \in E} \Big[\overline{f(\iota e)} f(\tau e) + \overline{f(\tau e)} f(\iota e) \Big]$$

$$= 2\,\mathrm{Re} \sum_{e \in E} \overline{f(\iota e)} f(\tau e),$$

in particular

$$\langle f, Af \rangle = 2 \sum_{e \in E} f(\iota e) f(\tau e) \text{ for } f \in \mathbb{R}^X. \tag{2.14}$$

Theorem 2.4.1 (Largest A-eigenvalue) *Let A be the adjacency matrix of a graph G, then the following statements hold:*

(a) the eigenvalue $\lambda_{\max}(A)$ has the largest absolute value among all eigenvalues of A.

(b) If G is connected, then the eigenvalue $\lambda_{\max}(A)$ is simple with a positive eigenvector.

(c) It holds

    (i) $\lambda_{\min}(A) < 0 < \lambda_{\max}(A)$, if G has at least one edge,

    (ii) $\lambda_{\min}(A) = \lambda_{\max}(A) = 0$, if G has no edges.

Proof The matrix A satisfies the conditions of the Perron-Frobenius theorem (for the weak version: always, for the strong version: if G is connected), so (a) follows from Theorem 2.3.2 and (b) follows from Theorem 2.3.3.

(c) If G has no edges, then $A = 0$ and all eigenvalues are zero. If there is at least one edge, then

$$\langle \mathbb{1}, A\mathbb{1} \rangle = \sum_{x,y \in X} A_{x,y} > 0, \quad \lambda_{\max}(A) \geq \frac{\langle \mathbb{1}, A\mathbb{1} \rangle}{\langle \mathbb{1}, \mathbb{1} \rangle} > 0.$$

Since the sum of all eigenvalues of A is zero (trace theorem), the smallest eigenvalue must be negative. $\qquad\square$

Now we discuss the behavior of the A-eigenvalues when removing vertices or edges.

Lemma 2.4.2 (Vertex Monotonicity for A) *Let $G = (X, E)$ be a graph with $|X| \geq 2$ and let A be its adjacency matrix. For every vertex $x_0 \in X$, the adjacency matrix A_0 of $G \setminus x_0$ satisfies*

$$\lambda_{\min}(A) \leq \lambda_{\min}(A_0) \leq \lambda_{\max}(A_0) \leq \lambda_{\max}(A).$$

Proof The matrix A_0 is obtained by deleting the x_0-row and the x_0-column in A, so the statement follows from the interlacing principle for matrices (Corollary 2.1.9).
$\qquad\square$

Lemma 2.4.3 (Edge Monotonicity for A) *Let $G = (X, E)$ be a graph. For every edge $e_0 \in E$, the adjacency matrix A_0 of $G \setminus e_0$ satisfies*

$$\lambda_{\max}(A_0) \leq \lambda_{\max}(A).$$

Proof Let $f \geq 0$ be an eigenvector of A_0 to $\lambda_{\max}(A_0)$. Using the representation (2.14) we obtain

$$\langle f, Af \rangle = \langle f, A_0 f \rangle + 2 f(\iota e_0) f(\tau e_0)$$

$$\equiv \lambda_{\max}(A_0)\|f\|^2 + 2 \underbrace{f(\iota e_0) f(\tau e_0)}_{\geq 0} \geq \lambda_{\max}(A_0)\|f\|^2,$$

and it follows

$$\lambda_{\max}(A) \geq \frac{\langle f, Af \rangle}{\|f\|^2} \geq \lambda_{\max}(A_0).$$

$\qquad\square$

Remark 2.4.4 One can also show (Exercise 2.4.2) that for <u>connected</u> G in the two previous lemmas the strict inequality $\lambda_{\max}(A_0) < \lambda_{\max}(A)$ holds.

Corollary 2.4.5 (Subgraphs and A-eigenvalues) *Let G be a graph and G' its subgraph, then for their adjacency matrices A and A' it holds:*

$$\lambda_{\max}(A) \geq \lambda_{\max}(A').$$

Proof Every subgraph of G is obtained by removing edges and vertices and at each step $\lambda_{\max}$ can only decrease (Lemmas 2.4.2 and 2.4.3).
$\qquad\square$

Recall that for any $f \in \mathbb{K}^X$ we have for the matrices L and Q of G:

$$\langle f, Lf \rangle = \sum_{e \in E} \left| f(\iota e) - f(\tau e) \right|^2, \qquad \langle f, Qf \rangle = \sum_{e \in E} \left| f(\iota e) + f(\tau e) \right|^2.$$

A significant difference to A is that only non-negative summands occur in the two sums.

Theorem 2.4.6 (Edge Monotonicity for L and Q) *Let $G = (X, E)$ be a graph and let L and Q be the corresponding Laplace matrices. For each $e_0 \in E$, the Laplace matrices L_0 and Q_0 of $G_0 := G \setminus e_0$ satisfy*

$$\lambda_k(L_0) \le \lambda_k(L) \text{ and } \lambda_k(Q_0) \le \lambda_k(Q) \text{ for all } k \in \left\{ 1, \dots, |X| \right\}.$$

Proof All mappings L, L_0, Q, Q_0 are defined on the same vector space $\mathbb{K}^X$ and for each $f \in \mathbb{K}^X$ we have

$$\langle f, Lf \rangle = \langle f, L_0 f \rangle + \left| f(\iota e_0) - f(\tau e_0) \right|^2 \ge \langle f, L_0 f \rangle,$$

$$\langle f, Qf \rangle = \langle f, Q_0 f \rangle + \left| f(\iota e_0) + f(\tau e_0) \right|^2 \ge \langle f, Q_0 f \rangle,$$

so the statement follows directly from the min-max formula (Corollary 2.1.7). $\square$

Lemma 2.4.7 (Vertex Monotonicity for L and Q) *Let $G = (X, E)$ be a graph with $|X| \ge 2$. For each $x_0 \in X$ and the graph $G_0 := G \setminus x_0$, the Laplace matrices L_0 and Q_0 of G_0 satisfy*

$$\lambda_{k-1}(L_0) \le \lambda_k(L) \text{ and } \lambda_{k-1}(Q_0) \le \lambda_k(Q) \text{ for all } k \in \left\{ 2, \dots, |X| \right\}.$$

Proof Let $H := \left(\{x_0\}, \varnothing \right)$ be the graph consisting only of the vertex x_0, then the corresponding Laplace matrices are simply 1×1 zero matrices with the only eigenvalue 0. As a reminder, all L and Q eigenvalues are always non-negative.

Consider $\widetilde{G} := G \setminus \{e \in E : e \sim x_0\}$ and the corresponding Laplace matrices $\widetilde{L}$ and $\widetilde{Q}$, then

$$\widetilde{G} = G_0 \sqcup H, \quad \widetilde{L} = L_0 \oplus 0_{\mathbb{K}}, \quad \widetilde{Q} = Q_0 \oplus 0_{\mathbb{K}},$$

$$\lambda_k(\widetilde{L}) = \lambda_{k-1}(L_0) \text{ and } \lambda_k(\widetilde{Q}) = \lambda_{k-1}(Q_0) \text{ for all } k \ge 2.$$

Moreover, by edge monotonicity (Lemma 2.4.6)

$$\lambda_k(\widetilde{L}) \le \lambda_k(L), \quad \lambda_k(\widetilde{Q}) \le \lambda_k(Q) \text{ for all } k \in \left\{ 1, \dots, |X| \right\},$$

and by combining the two inequalities, we obtain the statement. $\square$

Corollary 2.4.8 (Subgraphs and L- and Q-eigenvalues) *If G' is a subgraph of G, then for the corresponding Laplace matrices L' and L we have*

$$\lambda_{\max}(L') \le \lambda_{\max}(L), \quad \lambda_{\max}(Q') \le \lambda_{\max}(Q).$$

Proof We obtain G' by removing vertices and edges in G, where at each step the largest eigenvalue can only decrease (Lemmas 2.4.6 and 2.4.7). $\qquad\square$

Remark 2.4.9 From edge monotonicity, it follows that the complete graph K_n maximizes the value $\lambda_{\max}(A)$ among all connected graphs with n vertices. It is also known that the path graph W_n (see Exercise 1.6.4) minimizes this value in the same class of graphs [77]. Thus, for all connected graphs with n vertices, the inequality

$$2 \cos \frac{\pi}{n+1} \le \lambda_{\max}(A) \le n - 1.$$

holds.

Exercise 2.4.1 Let $G = (X, E)$ be a graph, $e \in E$ and $G' := G \setminus e$. Let $L := L_G$ and $L' := L_{G'}$ be the corresponding Laplace matrices. Show:

1. There exists a matrix N with $\dim \operatorname{ran} N = 1$ such that $L = L' + N$.
2. We have $\lambda_1(L') \le \lambda_1(L) \le \lambda_2(L') \le \lambda_2(L) \le \dots$.
   Hint: Use Exercise 2.1.1.
3. For at least one estimate, the inequality is strict.

Exercise 2.4.2 Let $G = (X, E)$ be a <u>connected</u> graph. Show:

1. For all $e \in E$, $\lambda_{\max}(A_{G\setminus e}) < \lambda_{\max}(A_G)$.
2. For all $x \in X$, $\lambda_{\max}(A_{G\setminus x}) < \lambda_{\max}(A_G)$.
   Hint: Let f be an eigenfunction for the largest eigenvalue of $A_{G\setminus x}$. Use its extension by 0 as a test function for A_G.

Exercise 2.4.3 From the article [40].

1. Calculate the largest Q-eigenvalue for the path graphs W_n (Exercise 1.6.4) as well as for the graphs $K_{1,3}$ and C_n.
   Hint: Exercise 1.6.1 may be helpful. The line graph of a path graph is again a path graph.
2. Deduce: For a graph G, the inequality $\lambda_{\max}(Q_G) < 4$ holds exactly when G is a disjoint union of paths and isolated vertices.

Exercise 2.4.4 Let $G = (X, E)$ be a connected graph. Show that the eigenfunctions for the largest A-eigenvalue are invariant with respect to all symmetries of G: If $\varphi : X \to X$ is a symmetry of G and f is an eigenfunction of A_G for $\lambda_{\max}(A_G)$, then $f(\varphi(x)) = f(x)$ for all $x \in X$.
   Hint: First show that for every eigenvalue μ of the linear mapping

$$\mathbb{K}^X \ni g \mapsto g \circ \varphi \in \mathbb{K}^X$$

we have $|\mu| = 1$.

Exercise 2.4.5 From the article [120].

Let $G = (X, E)$ be a connected graph and let A be its adjacency matrix. We want to show the inequality

$$\lambda_{\max}(A) \le \sqrt{2|E| - |X| + 1}. \tag{2.15}$$

Denote $\lambda := \lambda_{\max}(A)$ and let f be the corresponding positive and normalized eigenvector. For $x \in X$ define

$$f_x : X \to \mathbb{R}, \quad f_x(y) = \begin{cases} f(y), & y \sim x, \\ 0 & \text{otherwise,} \end{cases} \qquad A_x : V \to \mathbb{R}, \quad A_x(y) = A_{x,y}.$$

1. Show that for all $x \in X$ it holds

$$\lambda f(x) = \langle A_x, f_x \rangle, \quad \left(\lambda f(x)\right)^2 \le \deg x \left(1 - \sum_{y \not\sim x} f(y)^2\right).$$

2. Deduce:

$$\lambda^2 \le 2|E| - \sum_{x \in X} \deg x \sum_{y \not\sim x} f(y)^2.$$

3. Show the inequality (2.15).
4. Find all graphs for which equality in (2.15) is achieved.

Exercise 2.4.6 From the article [67].

Let $n \in \mathbb{N}$ and let $H_n = (X, E)$ be the hypercube defined in Exercise 1.6.3. Let $Y \subset X$ with $|Y| = 2^{n-1} + 1$. We want to show that for the subgraph of H_n induced by Y it holds:

$$\deg_{\max}(H_n)_Y \ge \sqrt{n}.$$

1. Show: There exists $Y \subset X$ with $|Y| = 2^n$ such that the induced graph $(H_n)_Y$ only has isolated vertices.
2. Let the following recursion be given:

$$A_1 = \begin{pmatrix} 0 & 1 \\ 1 & 0 \end{pmatrix}, \quad A_n = \begin{pmatrix} A_{n-1} & I \\ I & -A_{n-1} \end{pmatrix} \quad \text{for } n \ge 2.$$

Show: $\sigma(A_n) = \left\{ -\sqrt{n} \text{ with multiplicity } 2^{n-1}, \sqrt{n} \text{ with multiplicity } 2^{n-1} \right\}$.

3. Let $G = (X, E)$ be an arbitrary graph and B a Hermitian $X \times X$ matrix with

$$|B_{x,y}| \leq 1 \text{ for all } (x, y), \qquad B_{x,y} = 0 \text{ for } \{x, y\} \notin E.$$

Show: $\lambda_{\max}(B) \leq \deg_{\max} G$.

4. Let now $G = (H_n)_Y$.

   (a) Let B be the matrix obtained by deleting all rows and columns with numbers in $X \setminus Y$ in the matrix A_n. Show: $\lambda_{\max}(B) \geq \sqrt{n}$.
   Hint: Use the interlacing theorem.

   (b) Derive: $\deg_{\max} G \geq \sqrt{n}$.

Exercise 2.4.7 (Construction of Paths Using Eigenvectors) From [102].

Let $G = (X, E)$ be a connected graph and L its Laplace matrix. Let further $x_0 \in X$ such that $G \setminus x_0$ is still connected. We denote by L' the matrix obtained by deleting the x_0 row and the x_0 column in L, and consider

$$\lambda' := \min_{g \in \mathbb{R}^X,\ g(x_0)=0,\ g \neq 0} \frac{\langle g, Lg \rangle}{\|g\|^2}.$$

Show:

1. $\lambda' > 0$.
2. λ' is the smallest eigenvalue of L'.
3. λ' is a simple eigenvalue of L', and every real eigenvector of L' to λ' has constant sign and no zeros.

Let now $f' \in \mathbb{R}^{X \setminus \{x_0\}}$ be an eigenvector of L' with eigenvalue λ' with $f' > 0$. Define $f \in \mathbb{R}^X$ as the extension of f' by 0 on X.

4. Show that for every $x \neq x_0$ the iterative movement to a neighboring vertex y with $f(y) < f(x)$ leads to x_0 in a finite number of steps.

2.5 Bipartite Graphs

We continue the discussion of bipartite graphs started in Proposition 1.6.3.

Theorem 2.5.1 (Spectral Characterization of Connected Bipartite Graphs) *A connected graph G is bipartite if and only if for its adjacency matrix A it holds*

$$\lambda_{\min}(A) = -\lambda_{\max}(A). \tag{2.16}$$

Proof If G is bipartite, then $\sigma(A)$ is symmetric with respect to 0 (Proposition 1.6.3) and the statement is true.

Assume now that (2.16) is fulfilled, then $\lambda_{\max} = |\lambda_{\min}|$. The case $\lambda_{\max} = \lambda_{\min} = 0$ corresponds to $A = (0)$, i.e., to the graph with a single vertex, which is obviously bipartite (one partition component is empty). Now assume that $\lambda_{\min} < 0 < \lambda_{\max}$.

Let f be an eigenvector of A to $\lambda_{\min}$. We can assume $f \in \mathbb{R}^X$ (otherwise we take $\mathrm{Re}\, f$ or $\mathrm{Im}\, f$ instead of f). Then

$$
\lambda_{\max}\big\|\,|f|\,\big\|^2 = |\lambda_{\min}|\,\|f\|^2 = \big|\lambda_{\min}\|f\|^2\big| = \big|\langle f, Af\rangle\big|
$$

$$
= \Big|2\sum_{e\in E} f(\iota e)f(\tau e)\Big| \le 2\sum_{e\in E}\big|f(\iota e)\big|\,\big|f(\tau e)\big| \tag{2.17}
$$

$$
= \langle |f|, A|f|\rangle \le \lambda_{\max}\big\|\,|f|\,\big\|^2
$$

so $\langle |f|, A|f|\rangle = \lambda_{\max}\big\|\,|f|\,\big\|^2$ and it follows $|f| \in \ker(A - \lambda_{\max}I)$. Since the subspace $\ker(A - \lambda_{\max}I)$ is one-dimensional and spanned by a positive eigenvector (Theorem 2.4.1), the vector $|f|$ is a multiple of a positive eigenvector. Since $|f| \not\equiv 0$ we obtain $|f(x)| > 0$ for all $x \in X$. It follows that the disjoint sets

$$
Y := \big\{x \in X : f(x) > 0\big\}, \quad Y' := \big\{x \in X : f(x) < 0\big\},
$$

exhaust the vertex set X: $X = Y \cup Y'$. Since f and $|f|$ are eigenvectors to two different eigenvalues, $\langle f, |f|\rangle = 0$. Since $|f| > 0$, f cannot have a constant sign, so both Y and Y' are non-empty. Then there is an edge e' between Y and Y', and $f(\iota e')f(\tau e') < 0$.

Since the left-hand side and the right-hand side of (2.17) are equal, all inequalities in the chain are equalities. This is only possible if all terms in the first sum in (2.17) have the same sign. Since f is not zero at any vertex, the products $f(\iota e)f(\tau e)$ are either all positive or all negative. We have just seen that the product for $e := e'$ is negative, so they are all negative. It follows that every edge goes from Y to Y' and G is thus bipartite. $\qquad\square$

Corollary 2.5.2 (Spectral Characterization of Bipartite Graphs) *A graph G is bipartite if and only if for its adjacency matrix A it holds*

$$
\sigma(A) = -\sigma(A). \tag{2.18}
$$

Remark 2.5.3 The equality (2.18) is to be understood in the sense of multisets. In detail, (2.18) can be written in the following form:

$$
\dim \ker(A - \lambda I) = \dim \ker(A + \lambda I) \text{ for all } \lambda \in \mathbb{R}.
$$

Proof of Corollary 2.5.2 If G is bipartite, then (2.18) is fulfilled (Proposition 1.6.3). Conversely, assume now (2.18) is fulfilled, then

$$\lambda_{\max}(A) = -\lambda_{\min}(A) \equiv \left|\lambda_{\min}(A)\right|. \tag{2.19}$$

We perform induction on the number of connected components of G. If G has only one connected component, then G is connected, and thanks to (2.19) the graph G is bipartite (Theorem 2.5.1).

Let now $N \geq 2$, such that all graphs with at most $N - 1$ connected components, whose adjacency matrices A' fulfill the condition $\sigma(A') = -\sigma(A')$, are bipartite. Let G be a graph with N connected components $G_1, \ldots, G_N$ with adjacency matrices $A_1, \ldots, A_N$, then for the adjacency matrix A of G one has

$$A = A_1 \oplus \cdots \oplus A_N, \quad \sigma(A) = \sigma(A_1) \cup \cdots \cup \sigma(A_N).$$

There is a connected component G_k with $\lambda_{\min}(A) \in \sigma(A_k)$. Then we have $\lambda_{\min}(A) = \lambda_{\min}(A_k)$, in addition one has the general inequalities $\lambda_{\max}(A_k) \leq \lambda_{\max}(A)$ and $\left|\lambda_{\min}(A_k)\right| \leq \lambda_{\max}(A_k)$. Overall

$$\left|\lambda_{\min}(A)\right| = \left|\lambda_{\min}(A_k)\right| \leq \lambda_{\max}(A_k) \leq \lambda_{\max}(A) = \left|\lambda_{\min}(A)\right|,$$

and it follows that $\left|\lambda_{\min}(A_k)\right| = \lambda_{\max}(A_k)$. According to theorem 2.5.1, G_k is a bipartite graph, so $\sigma(A_k) = -\sigma(A_k)$.

We remove G_k and consider the graph

$$G' := \bigsqcup_{j:\, j \neq k} G_j \equiv G \setminus G_k.$$

The adjacency matrix A' of G' is $A' = \bigoplus_{j:\, j \neq k} A_j$, and we have

$$A = A' \oplus A_k, \quad \sigma(A') = \sigma(A) \setminus \sigma(A_k).$$

Because of $\sigma(A) = -\sigma(A)$ and $\sigma(A_k) = -\sigma(A_k)$, it also holds $\sigma(A') = -\sigma(A')$. Since G' only has $N - 1$ connected components, G' is bipartite according to the induction assumption. Then $G \equiv G' \sqcup G_k$ is also bipartite. $\qquad\square$

Remark 2.5.4 The symmetry of the spectrum of the adjacency matrix for bipartite graphs is one of the first results of spectral graph theory, see e.g. [33, Chapter 3a]. The converse (Theorem 2.5.1) was found in 1969 by Cvetković [35].[1]

Remark 2.5.5 (Bipartiteness and L-eigenvalues) The bipartiteness cannot be read off from the eigenvalues of the Laplace matrix L. The Laplace matrices of

---

[1] See also historical comments in [38, after Theorem 3.11].

Fig. 2.1 Non-isomorphic graphs with the same L-spectra

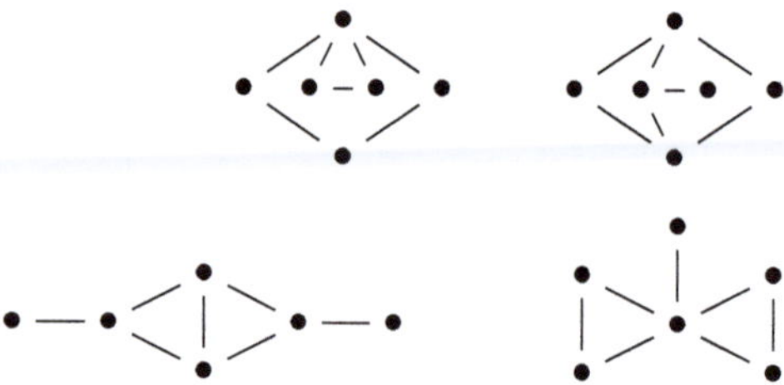

Fig. 2.2 Non-isomorphic graphs with the same A-spectra

the graphs in Fig. 2.1 have the same characteristic polynomial

$$\lambda \mapsto \lambda(\lambda - 3)^2(\lambda - 2)(\lambda^2 - 6\lambda + 4)$$

and thus these graphs have the same L-eigenvalues. However, only the second graph is bipartite.

Remark 2.5.6 The entire structure of the graph is not uniquely determined by the A-eigenvalues. The adjacency matrices of the graphs in Fig. 2.2 have the same characteristic polynomial

$$\lambda \mapsto (\lambda^2 - 1)(\lambda^4 - 6\lambda^2 - 4\lambda + 1)$$

and thus the two graphs have the same A-eigenvalues, even though they are not isomorphic to each other.

Exercise 2.5.1 Let $G = (X, E)$ be a graph and L and Q its Laplace matrices. Show the equivalence of the following three statements:

1. G is bipartite,
2. $\sigma(L) = \sigma(Q)$,
3. $\dim \ker L = \dim \ker Q$.

Exercise 2.5.2 Let $G = (X, E)$, with $|X| = n$, be a connected graph that is <u>not</u> bipartite. Furthermore, let A be its adjacency matrix with the eigenpairs $(\lambda_1, f_1), \ldots, (\lambda_{\max} = \lambda_n, f_n)$, such that $f_1, \ldots, f_n$ form an orthonormal basis of $\mathbb{R}^X$. For $k \in \mathbb{N}$ define $N_k, s_k : X \to \mathbb{R}$ by

$$N_k(x) := \text{The number of edge paths of length } k \text{ that start in } x,$$

$$s_k(x) := \frac{N_k(x)}{\langle \mathbb{1}, N_k \rangle}.$$

We want to show that s_k converges for $k \to \infty$ to an eigenvector of A for the largest eigenvalue λ_n. Show:

1. $N_k = A^k \mathbb{1}$,
2. Use the expansion in the basis $f_1, \ldots, f_n$ to show that

$$\lim_{k \to \infty} s_k = \frac{f_n}{\langle \mathbb{1}, f_n \rangle}.$$

Exercise 2.5.3 (Graham-Pollak-Witsenhausen Theorem) From the articles [57, 58].

Let $G = (X, E)$ be a graph, with $n := |X|$. We say that subgraphs $H_i = (X_i, E_i)$ of G, with $i \in \{1, \ldots, p\}$, form a *complete bipartite cover* of G if the following conditions are met:

- $E_1, \ldots, E_p$ are pairwise disjoint with $E_1 \cup \ldots \cup E_p = E$,
- $X_1 \cup \ldots \cup X_p = X$,
- all H_i are complete bipartite graphs.

The *bipartite dimension $p(G)$* of G is defined as the minimal p such that a complete bipartite cover of G with p subgraphs exists.

Let $p := p(G)$ and let $H_1 \ldots, H_p$ be as above with adjacency matrices B_i. Also consider the graphs $G_i := (X, E_i)$ and their adjacency matrices A_i. Show:

1. The adjacency matrix of G is $A := A_1 + \cdots + A_p$.
2. For each $i \in \{1, \ldots, p\}$ there are subspaces $U_i^+, U_i^- \subset \mathbb{K}^X$ with $\dim U_i^+ = \dim U_i^- = n - 1$ and

$$\langle u, A_i u \rangle \geq 0 \text{ for all } u \in U_i^+, \qquad \langle u, A_i u \rangle \leq 0 \text{ for all } u \in U_i^-.$$

Hint: Use the calculations from example 1.6.7.
3. There are $(n - p)$-dimensional subspaces $U_+, U_- \subset \mathbb{K}^X$, such that

$$\langle u, A u \rangle \geq 0 \text{ for all } u \in U_+, \qquad \langle u, A u \rangle \leq 0 \text{ for all } u \in U_-.$$

4. The inequalities $\lambda_{n-p}(A) \leq 0$ and $\lambda_{p+1}(A) \geq 0$ hold.
5. The inequality

$$p(G) \geq \max \big\{ \text{the number of positive eigenvalues of } A,$$
$$\text{the number of negative eigenvalues of } A \big\}.$$

6. $p(K_n) = n - 1$.

Hint: First show $p(K_n) \geq n - 1$, then construct a specific complete bipartite cover with $n - 1$ subgraphs.

2.6 Chromatic Number

In this section we prove upper and lower estimates for the chromatic number $\chi(G)$ of a graph G using its A-eigenvalues. We start with a basic inequality.

Lemma 2.6.1 (A-eigenvalues and Degrees) *For the adjacency matrix A of a graph $G = (X, E)$, the inequalities*

$$\frac{\sum_{x \in X} \deg x}{|X|} =: \deg_{\text{avg}} G \leq \lambda_{\max}(A) \leq \deg_{\max} G := \max_{x \in X} \deg x. \qquad (2.20)$$

hold.

Proof Let f be an eigenvector of A for $\lambda_{\max}$. Let $y \in X$ with

$$|f(y)| = \max_{x \in X} |f(x)|,$$

then $|f(y)| > 0$. We have

$$|\lambda_{\max}| |f(y)| = |\lambda_{\max} f(y)| = |(Af)(y)|$$

$$= \left| \sum_{x:x \sim y} f(x) \right| \leq \sum_{x:x \sim y} |f(y)| = (\deg y)|f(y)|,$$

so $|\lambda_{\max}| \leq \deg y \leq \deg_{\max} G$. This proves the right-hand side of (2.20). For the left-hand side we use

$$\langle \mathbb{1}, \mathbb{1} \rangle = |X|, \quad \langle \mathbb{1}, A\mathbb{1} \rangle = \sum_{x \in X} \deg x, \quad \lambda_{\max} \geq \frac{\langle \mathbb{1}, A\mathbb{1} \rangle}{\langle \mathbb{1}, \mathbb{1} \rangle} = \deg_{\text{avg}} G.$$

$\square$

Although the inequality (2.20) may seem a bit naive, it provides good results in practice. In particular, it leads to the following estimate for the chromatic number, which was found by Wilf in 1967 [114].

Theorem 2.6.2 (Wilf: Upper Estimate for the Chromatic Number) *Let G be a graph and A its adjacency matrix, then*

$$\chi(G) \leq \lambda_{\max}(A) + 1. \qquad (2.21)$$

Proof Since $\chi(G)$ is an integer, (2.21) is equivalent to

$$\chi(G) \le \big[\lambda_{\max}(A)\big] + 1, \tag{2.22}$$

where $[a] := \max\{m \in \mathbb{Z} : m \le a\}$ is the integer part of a.

We use induction on $n :=$ the number of vertices in G. For $n = 1$ one has $\chi(G) = 1$ and $A = (0)$ with $\lambda_{\max}(A) = 0$, so (2.22) is satisfied in this case.

Let now $n \ge 2$ such that (2.22) is satisfied for all graphs with at most $n - 1$ vertices and let $G = (X, E)$ be a graph with n vertices and adjacency matrix A. By Lemma 2.6.1, $\lambda_{\max}(A) \ge \deg_{\text{avg}} G$. There is at least one vertex $x \in X$ with $\deg x \le \deg_{\text{avg}} G$, then also $\deg x \le \lambda_{\max}(A)$ and thus $\deg x \le \big[\lambda_{\max}(A)\big]$.

Consider the graph $G' := G \setminus x$ and its adjacency matrix A'. Since G' has only $n - 1$ vertices, by the induction hypothesis

$$\chi(G') \le \big[\lambda_{\max}(A')\big] + 1 \le \big[\lambda_{\max}(A)\big] + 1$$

(where the second inequality holds due to the vertex monotonicity for A). Thus, there is a vertex coloring of G' with $k := \big[\lambda_{\max}(A)\big] + 1$ colors.

Thanks to the special choice, x has at most $k - 1$ neighbors in G, which occupy at most $k - 1$ of the k available colors, so there is a free color for x. Thus, we have constructed a vertex coloring for G with k colors. $\square$

Example 2.6.3 We use the calculations from Sect. 1.6.

(a) For the complete graph K_n, $\chi(K_n) = n$ and $\lambda_{\max}(A) = n - 1$, and the estimate (2.21) is optimal.
(b) For the cycles C_n,

$$\chi(C_n) = \begin{cases} 2, & n \text{ even}, \\ 3, & n \text{ odd}, \end{cases} \qquad \lambda_{\max}(A) = \max_{k \in \{0,\dots,n-1\}} 2\cos\frac{2\pi k}{n} = 2,$$

so the estimate (2.21) is optimal for odd n.
(c) For the complete bipartite graph $K_{1,n}$ (a star with one central vertex and n rays), $\chi(K_{1,n}) = 2$, $\lambda_{\max}(A) = \sqrt{n}$, and for large n the upper estimate (2.21) is far from the optimal value.

In 1970, Hoffman [65] also discovered an important lower estimate:

Theorem 2.6.4 (Hoffman: Lower Estimate for the Chromatic Number) *Let G be a graph with at least one edge and A its adjacency matrix, then*

$$\chi(G) \ge 1 + \frac{\lambda_{\max}(A)}{-\lambda_{\min}(A)}. \tag{2.23}$$

Proof Denote $k := \chi(G)$. Let $c : X \to \{1, \ldots, k\}$ be a vertex coloring for G and consider the sets

$$X_j := \{x \in X : c(x) = j\} \subset X, \quad j \in \{1, \ldots, k\}.$$

Thanks to the choice of k, all X_j are non-empty. We also consider the subspaces

$$F_j := \{f \in \mathbb{K}^X : f(x) = 0 \text{ for all } x \notin X_j\} \subset \mathbb{K}^X.$$

We have the orthogonal decomposition $\mathbb{K}^X = F_1 \oplus \cdots \oplus F_k$.

Let $f_j \in F_j$. For each edge $e \in E$, at least one of the end vertices $\iota e, \tau e$ is <u>not</u> in X_j, so $\overline{f_j(\iota e)} f_j(\tau e) = 0$, and it follows

$$\langle f_j, A f_j \rangle = 2 \sum_{e \in E} \operatorname{Re} \overline{f_j(\iota e)} f_j(\tau e) = 0 \text{ for all } f_j \in F_j. \tag{2.24}$$

Let f be an eigenvector of A corresponding to $\lambda_{\max}(A)$, then f can be represented as

$$f = \sum_{j=1}^{k} a_j u_j, \quad a_j \in \mathbb{K}, \quad u_j \in F_j, \tag{2.25}$$

where we assume w.l.o.g. that $\|u_j\| = 1$ for all j. Now consider the subspace $U := \operatorname{span}\{u_1, \ldots, u_k\} \subset \mathbb{K}^X$ and the natural embedding

$$\Theta : U \ni g \mapsto g \in \mathbb{K}^X.$$

The adjoint mapping $\Theta^* : \mathbb{K}^X \to U$ is the orthogonal projection onto U and $\Theta^* \Theta : U \to U$ is the identity mapping. By the interlacing principle (Corollary 2.1.8), $\lambda_{\max}(\Theta^* A \Theta) \leq \lambda_{\max}(A)$ holds. From (2.25) it follows that $f \in U$ and

$$\|f\|_U = \|f\|_{\mathbb{K}^X}, \quad \langle f, \Theta^* A \Theta f \rangle_U = \langle \Theta f, A \Theta f \rangle_{\mathbb{K}^X} = \langle f, A f \rangle_{\mathbb{K}^X} = \lambda_{\max}(A) \|f\|_{\mathbb{K}^X}^2,$$

$$\lambda_{\max}(\Theta^* A \Theta) \geq \frac{\langle f, \Theta^* A \Theta f \rangle_U}{\|f\|_U^2} \equiv \frac{\lambda_{\max}(A) \|f\|_{\mathbb{K}^X}^2}{\|f\|_{\mathbb{K}^X}^2} = \lambda_{\max}(A),$$

so overall the equality

$$\lambda_{\max}(\Theta^* A \Theta) = \lambda_{\max}(A). \tag{2.26}$$

According to the above construction, $(u_1, \ldots, u_k)$ is an orthonormal basis in U. For the representation matrix $M = (M_{i,j})_{i,j \in \{1,\ldots,k\}}$ of $\Theta^* A \Theta$ in this basis, it holds

$$M_{i,j} = \langle u_i, \Theta^* A \Theta u_j \rangle = \langle \Theta u_i, A \Theta u_j \rangle = \langle u_i, A u_j \rangle,$$

and thanks to (2.24), $M_{j,j} = 0$ for all j. Since $\Theta^* A \Theta$ and M have the same eigenvalues, one can use the trace theorem for M:

$$\sum_{j=1}^{k} \lambda_j(\Theta^* A \Theta) \equiv \sum_{j=1}^{k} \lambda_j(M) = \sum_{j=1}^{k} \underbrace{M_{j,j}}_{=0} = 0.$$

There are a total of k eigenvalues and one can estimate the $k-1$ smallest eigenvalues by the smallest one, so it follows

$$(k-1)\lambda_{\min}(\Theta^* A \Theta) + \lambda_{\max}(\Theta^* A \Theta) \le 0.$$

We have the equality (2.26) and $\lambda_{\min}(A) \le \lambda_{\min}(\Theta^* A \Theta)$ (interlacing principle), so

$$(k-1)\lambda_{\min}(A) + \lambda_{\max}(A) \le 0.$$

Since G has at least one edge, $\lambda_{\min}(A) < 0 < \lambda_{\max}(A)$ and it follows

$$\chi(G) \equiv k \ge 1 + \frac{\lambda_{\max}(A)}{-\lambda_{\min}(A)}.$$

$\square$

Remark 2.6.5

(a) If G has no edges, the right-hand side of (2.23) is not defined, since $\lambda_{\min}(A) = 0$.

(b) Because of $\lambda_{\max}(A) \ge -\lambda_{\min}(A)$, the right-hand side of (2.23) is always ≥ 2. For bipartite graphs, the right-hand side is always 2: In this case, the estimate is optimal. For all non-bipartite connected graphs, the right-hand side is strictly greater than 2 (thanks to the spectral characterization in Theorem 2.5.1) and the Hoffman estimate states $\chi(G) \ge 3$.

(c) For the cycles C_n with odd n, the Wilf estimate and the Hoffman estimate give the exact value of $\chi(C_n)$, while for even n only the inequalities $2 \le \chi(C_n) \le 3$ are obtained.

(d) For K_n we have $\lambda_{\min}(A) = -1$ and $\lambda_{\max}(A) = n - 1$, so with both estimates the exact value $\chi(K_n) = n$ is obtained.

(e) In Remark 2.8.5 we will derive an additional estimate for $\chi(G)$ that takes into account $\lambda_{\max}(A)$ and $|X|$. There are further inequalities for the chromatic number that also use other eigenvalues (not just the smallest and the largest) of the adjacency matrix, see e.g. [61, 117].

Exercise 2.6.1 Show that a graph G with n vertices and adjacency matrix A is regular if and only if

$$\sum_{j=1}^{n} \lambda_j(A)^2 = n\lambda_n(A).$$

Hint: You can combine Corollary 1.5.14 (Trace theorem for adjacency matrices), Exercise 1.6.6 (Characterization of regular graphs) and the proof of Lemma 2.6.1.

2.7 Vertex Connectivity

Now we will examine the second smallest eigenvalue $\lambda_2(L)$ of the Laplace matrix L of a graph $G = (X, E)$. As a reminder, the vector $\mathbb{1}$ is always an eigenvector of L to $\lambda_1(L) \equiv 0$, so the following characterization follows from Corollary 2.1.6:

Corollary 2.7.1 *Let L be the Laplace matrix of a graph G, then*

$$\lambda_2(L) = \min_{\langle \mathbb{1}, f \rangle = 0,\, f \neq 0} \frac{\sum_{e \in E} \left| f(\iota e) - f(\tau e) \right|^2}{\sum_{x \in X} \left| f(x) \right|^2},$$

and $0 \neq f \in \mathbb{K}^X$ with $\langle \mathbb{1}, f \rangle = 0$ is an eigenvector to $\lambda_2(M)$ exactly when the minimum is reached on f.

We have already seen (Theorem 1.5.8) that $\lambda_2(L) > 0$ exactly when G is connected. Fiedler [50] discovered in 1973 that $\lambda_2(L)$ is closely related to the multiple connectivity: This was one of the strongest motivations for studying Laplace matrices.

Theorem 2.7.2 (Fiedler) *Let $G = (X, E)$ be a graph with Laplace matrix L. Let $Y \subsetneq X$ such that $G \setminus Y$ is <u>not</u> connected, then*

$$|Y| \geq \lambda_2(L).$$

We first need an intermediate step.

Lemma 2.7.3 *Let $G = (X, E)$ be a graph with $|X| \geq 3$ and let $x_0 \in X$, then for the Laplace matrices L and L_0 of G and $G_0 := G \setminus x_0$ one has*

$$\lambda_2(L_0) \geq \lambda_2(L) - 1.$$

Proof Let $n := |X|$. We construct a new graph $\widetilde{G}$, by connecting x_0 with all other vertices of G (if this was already the case, we set $\widetilde{G} := G$), then for its Laplace matrix $\widetilde{L}$ (thanks to edge monotonicity)

$$\lambda_2(\widetilde{L}) \geq \lambda_2(L). \tag{2.27}$$

We can assume w.l.o.g. that x_0 is the first vertex of X. In $\widetilde{G}$ the vertex x_0 has degree $n-1$, so the following block representation holds:

$$\widetilde{L} = \begin{pmatrix} n-1 & -\mathbb{1}^* \\ -\mathbb{1} & L_0 + I \end{pmatrix}, \quad \mathbb{1} := \begin{pmatrix} 1 \\ \cdot \\ \cdot \\ \cdot \\ \cdot \\ 1 \end{pmatrix} \Big\} \; n-1 \text{ rows.}$$

As a reminder: $\mathbb{1}$ is an eigenvector of L_0 to $\lambda_1(L_0) = 0$. Consider two cases:

- $\lambda_1(L_0) < \lambda_2(L_0)$. Let f_0 be any eigenvector of L_0 to $\lambda_2(L_0)$, then $\langle \mathbb{1}, f_0 \rangle = 0$.
- $\lambda_1(L_0) = \lambda_2(L_0)$. Then the eigenspace of L_0 to $\lambda_2(L_0)$ is at least two-dimensional and contains a vector $f_0 \neq 0$ with $\langle \mathbb{1}, f_0 \rangle = 0$.

So in both cases there is an eigenvector f_0 of L_0 to the eigenvalue $\lambda_2(L_0)$ with $\langle f_0, \mathbb{1} \rangle = 0$. With this we get

$$\widetilde{L} \begin{pmatrix} 0 \\ f_0 \end{pmatrix} = \begin{pmatrix} n-1 & -\mathbb{1}^* \\ -\mathbb{1} & L_0 + I \end{pmatrix} \begin{pmatrix} 0 \\ f_0 \end{pmatrix} = \begin{pmatrix} 0 \\ (L_0 + I) f_0 \end{pmatrix} = (\lambda_2(L_0) + 1) \begin{pmatrix} 0 \\ f_0 \end{pmatrix},$$

i.e., $\lambda_2(L_0) + 1 > 0$ is an eigenvalue of $\widetilde{L}$; the positivity follows from $\lambda_2(L_0) \geq 0$. We note that

$$\lambda_2(\widetilde{L}) \leq \lambda_k(\widetilde{L}) \text{ for all } k \text{ with } \lambda_k(\widetilde{L}) > 0,$$

and thus $\lambda_2(\widetilde{L}) \leq \lambda_2(L_0) + 1$. It follows $\lambda_2(L_0) \geq \lambda_2(\widetilde{L}) - 1 \overset{(2.27)}{\geq} \lambda_2(L) - 1.$ $\quad\square$

Proof of Theorem 2.7.2 (Fiedler) For the Laplace matrix L' of $G \setminus Y$ we have $\lambda_2(L') = 0$. By iteratively applying Lemma 2.7.3 we get

$$0 = \lambda_2(L') \geq \lambda_2(L) - |Y|,$$

so $|Y| \geq \lambda_2(L)$. $\quad\square$

Remark 2.7.4 If one uses the vertex connectivity $\kappa(G)$ and the edge connectivity $\kappa'(G)$ (see Sect. 1.3), one has the **Fiedler inequalities**

$$\lambda_2(L) \leq \kappa(G) \leq \kappa'(G), \text{ if } G \text{ is not a complete graph.}$$

The first inequality is exactly Fiedler's theorem and the second is Whitney's theorem. For the complete graph $G = K_n$, the first inequality is false: We have $\kappa(G) = n - 1 < n = \lambda_2(L)$. This is not a contradiction to Fiedler's theorem, as in this case no subset Y with the required properties exists. The number $\lambda_2(L)$ is often also called the **Fiedler value** or **algebraic connectivity** of G, and the eigenvectors of L corresponding to $\lambda_2(L)$ are often called **Fiedler vectors** of G.

Example 2.7.5 We again use the calculations from Sect. 1.6.

(a) For $G = K_{m,n}$ we have $\lambda_2(L) = \min\{m, n\}$. Here, the smaller of the two partition components forms a separating vertex set, so $\kappa(K_{m,n}) = \min\{m, n\} = \lambda_2(L)$ and the Fiedler estimate for $\kappa(K_{m,n})$ is optimal.
(b) Let $G = C_n$ with $n \geq 4$ (for $n = 3$, C_3 is a complete graph). We obviously have $\kappa(C_n) = 2$ and

$$\lambda_2(L) = 2 - 2\cos\frac{2\pi}{n} \in (1, 2]$$

(see Example 1.6.11), so the Fiedler inequality is optimal for $\kappa(C_n)$.

Remark 2.7.6 Fiedler [50] has also proven the lower estimate

$$2\kappa'(G)\left(1 - \cos\frac{2\pi}{|X|}\right) \leq \lambda_2(L)$$

for all graphs $G = (X, E)$ using advanced methods of matrix theory from [52]. The estimate is optimal for the path graphs W_n (see Exercise 1.6.4). In [17, Theorem 2.3] an alternative elementary proof was found using the min-max principle (the new proof includes a small detour via graphs with multiple edges and corresponding versions of Menger's theorem).

Exercise 2.7.1 From the article [44].
Let $G = (X, E)$ be a graph with signless Laplace matrix Q and let K be the minimum number of edges of G that must be removed to make the graph bipartite.
Let $H = (X', E')$ be a bipartite subgraph of G such that $|E'|$ is maximal. Let Y and Y' be the partition components of H.

1. Show: Either $X = X'$ or all vertices of $X \setminus X'$ are isolated in G.
2. Use the function $f := \mathbb{1}_Y - \mathbb{1}_{X \setminus Y}$ to show the inequality

$$K \geq \frac{\lambda_{\min}(Q)}{4}|X|.$$

2.8 Clique Number

Now we will find some spectral estimates for the clique number $\omega(G)$ (see Sect. 1.3). Most results are based on the following theorem published in 1965 [84].

Theorem 2.8.1 (Motzkin–Straus: Clique Number Through Maximization) *Let $G = (X, E)$ be a graph and A its adjacency matrix. Define*

$$\Lambda(G) := \left\{ f \in \mathbb{K}^X : f \geq 0 \text{ and } \sum_{x \in X} f(x) = 1 \right\} \subset \mathbb{K}^X,$$

$$F(G) := \max_{f \in \Lambda(G)} \langle f, Af \rangle.$$

Then

$$F(G) = \frac{\omega(G) - 1}{\omega(G)}. \tag{2.28}$$

We first prove some auxiliary statements about $F(G)$.

Lemma 2.8.2 *If $G = (X, E)$ is a complete graph, then*

$$F(G) = \frac{|X| - 1}{|X|}.$$

Proof Let $f \in \Lambda(G)$, then

$$\langle f, Af \rangle = \sum_{x \in X} f(x) \underbrace{\sum_{y \in X,\, y \neq x} f(y)}_{=1 - f(x)} = \sum_{x \in X} f(x)\big(1 - f(x)\big)$$

$$= \underbrace{\sum_{x \in X} f(x)}_{=1} - \underbrace{\sum_{x \in X} f(x)^2}_{=\|f\|^2} = 1 - \|f\|^2. \tag{2.29}$$

Thanks to the Cauchy-Schwarz inequality we have

$$1^2 = \left| \sum_{x \in X} f(x) \right|^2 \equiv |\langle \mathbb{1}, f \rangle|^2 \leq \|\mathbb{1}\|^2 \|f\|^2 \equiv |X| \, \|f\|^2,$$

so

$$\|f\|^2 \geq \frac{1}{|X|} \quad \text{and} \quad F(G) \leq 1 - \frac{1}{|X|} \equiv \frac{|X| - 1}{|X|}.$$

At the same time,

$$F(G) \geq \left\langle \frac{\mathbb{1}}{|X|}, A \frac{\mathbb{1}}{|X|} \right\rangle \stackrel{(2.29)}{\equiv} 1 - \left\| \frac{\mathbb{1}}{|X|} \right\|^2 \equiv 1 - \frac{1}{|X|} \equiv \frac{|X|-1}{|X|}.$$

$\square$

Lemma 2.8.3

(a) If G' is a subgraph of G, then $F(G') \leq F(G)$.

(b) Let $f \in \Lambda(G)$ with $F(G) = \langle f, Af \rangle$ and let $X' := \{x \in X : f(x) > 0\}$, then $F(G) = F(G_{X'})$, where $G_{X'}$ is the subgraph of G induced by X'.

Proof

(a) Let $G' = (X', E')$ and let A' be the corresponding adjacency matrix, then

$$A'_{x,y} \leq A_{x,y} \text{ for all } x, y \in X'.$$

Let $R : \mathbb{K}^{X'} \to \mathbb{K}^X$ be the extension by zero,

$$(Rg)(x) = \begin{cases} g(x), & x \in X', \\ 0, & x \in X \setminus X', \end{cases} \qquad g \in \mathbb{K}^{X'},$$

then $R\big(\Lambda(G')\big) \subset \Lambda(G)$. For each $g \in \Lambda(G')$, since $g \geq 0$,

$$\langle g, A'g \rangle = \sum_{x,y \in X'} A'_{x,y} g(x)g(y) \leq \sum_{x,y \in X'} A_{x,y} g(x)g(y)$$

$$= \sum_{x,y \in X'} A_{x,y}\big(Rg\big)(x)\big(Rg\big)(y) = \sum_{x,y \in X} A_{x,y}\big(Rg\big)(x)\big(Rg\big)(y)$$

$$= \langle Rg, ARg \rangle.$$

$$(2.30)$$

It follows

$$F(G') = \max_{g \in \Lambda(G')} \langle g, A'g \rangle \leq \max_{g \in \Lambda(G')} \langle Rg, ARg \rangle$$

$$= \max_{f \in R(\Lambda(G'))} \langle f, Af \rangle \leq \max_{f \in \Lambda(G)} \langle f, Af \rangle = F(G).$$

(b) Denote $G' := G_{X'}$ and define A' and R as in (a). In this case

$$A'_{x,y} = A_{x,y} \text{ for all } x, y \in X',$$

and as in (2.30) we get $\langle g, A'g \rangle = \langle Rg, ARg \rangle$ for all $g \in \Lambda(G')$. Let f' be the restriction of f to X', then $f = Rf'$, so

$$F(G) = \langle f, Af \rangle = \langle f', A'f' \rangle \leq F(G'),$$

where $F(G) \geq F(G')$ was already proven in (a).

$\square$

Proof of Motzkin-Straus Theorem 2.8.1 For $f \in \mathbb{K}^X$ define

$$\operatorname{supp} f := \{x \in X : f(x) \neq 0\}.$$

Among all functions $f \in \Lambda(G)$ with $F(G) = \langle f, Af \rangle$ we choose one with the smallest $|\operatorname{supp} f|$ and denote $X' := \operatorname{supp} f$. We want to prove the following auxiliary statement:

(H) The graph $G_{X'}$ induced by X' is complete.

Assume that (H) is false, then there are $y_1, y_2 \in X'$ with $y_1 \neq y_2$ and $y_1 \not\sim y_2$. W.l.o.g. we assume that

$$\sum_{x : x \sim y_1} f(x) \geq \sum_{x : x \sim y_2} f(x)$$

(otherwise swap y_1 and y_2). Define $g \in \mathbb{K}^X$ with

$$g(x) = \begin{cases} f(y_1) + f(y_2), & x = y_1, \\ 0, & x = y_2, \\ f(x), & x \in X \setminus \{y_1, y_2\}. \end{cases}$$

Clearly $g \in \Lambda(G)$, so $F(G) \geq \langle g, Ag \rangle$. We have (thanks to $y_1 \not\sim y_2$)

$$\langle f, Af \rangle = 2 \sum_{e \in E} f(\iota e) f(\tau e)$$

$$= 2 \sum_{e \sim y_1} f(\iota e) f(\tau e) + 2 \sum_{e \sim y_2} f(\iota e) f(\tau e) + 2 \sum_{e \not\sim y_1, e \not\sim y_2} f(\iota e) f(\tau e)$$

$$= 2 \sum_{x \sim y_1} f(y_1) f(x) + 2 \sum_{x \sim y_2} f(y_2) f(x) + 2 \sum_{e \not\sim y_1, e \not\sim y_2} f(\iota e) f(\tau e)$$

and

$$\langle g, Ag \rangle = 2 \sum_{x \sim y_1} \underbrace{g(y_1)}_{f(y_1)+f(y_2)} \underbrace{g(x)}_{f(x)} + 2 \sum_{x \sim y_2} \underbrace{g(y_2)}_{=0} \underbrace{g(x)}_{f(x)} + 2 \sum_{e \not\ni y_1,\, e \not\ni y_2} \underbrace{g(\iota e)}_{f(\iota e)} \underbrace{g(\tau e)}_{f(\tau e)}$$

$$= 2 \sum_{x \sim y_1} \big(f(y_1) + f(y_2)\big) f(x) + 2 \sum_{e \not\ni y_1,\, e \not\ni y_2} f(\iota e) f(\tau e),$$

so

$$\langle g, Ag \rangle - \langle f, Af \rangle = 2 \underbrace{f(y_2)}_{>0} \underbrace{\bigg(\sum_{x \sim y_1} f(x) - \sum_{x \sim y_2} f(x) \bigg)}_{\geq 0} \geq 0,$$

and $\langle g, Ag \rangle \geq \langle f, Af \rangle = F(G)$. It follows $\langle g, Ag \rangle = F(G)$. But

$$| \operatorname{supp} g | = \big|(\operatorname{supp} f) \setminus \{y_2\}\big| = | \operatorname{supp} f | - 1 < | \operatorname{supp} f |,$$

which contradicts the choice of f. This proves (H).

It follows that X' is a clique in G, so $\omega(G) \geq |X'|$. According to Lemmas 2.8.2 and 2.8.3(b),

$$F(G) = F(G_{X'}) = \frac{|X'| - 1}{|X'|}.$$

Let $Y \subset X$ be a clique with $|Y| = \omega(G)$. Thanks to Lemma 2.8.3(a),

$$\frac{|X'| - 1}{|X'|} = F(G) \geq F(G_Y) \equiv \frac{|Y| - 1}{|Y|} \equiv \frac{\omega(G) - 1}{\omega(G)},$$

and it follows that $|X'| \geq \omega(G)$. Since the reverse inequality was also proven above, we have $|X'| = \omega(G)$. $\square$

Now, the largest A-eigenvalue can be integrated into the inequalities [115].

Corollary 2.8.4 (Wilf: Inequalities for the Clique Number) *Let $G = (X, E)$ be a graph with adjacency matrix A and clique number $\omega(G)$, then*

$$\omega(G) - 1 \leq \lambda_{\max}(A) \leq \frac{\omega(G) - 1}{\omega(G)} |X| \tag{2.31}$$

or equivalently

$$\frac{|X|}{|X| - \lambda_{\max}(A)} \leq \omega(G) \leq 1 + \lambda_{\max}(A). \tag{2.32}$$

Proof Let $Y \subset X$ be a clique with $|Y| = \omega(G)$ and let A_Y be the adjacency matrix of the induced graph G_Y. The largest eigenvalue of A_Y is $|Y| - 1 \equiv \omega(G) - 1$ (Example 1.6.2) and thanks to the monotonicity, we get

$$\lambda_{\max}(A) \geq \lambda_{\max}(A_Y) = \omega(G) - 1.$$

Alternatively, the comparison with the chromatic number can be used: thanks to Proposition 1.3.4 and Wilf's theorem (Theorem 2.6.2),

$$\omega(G) \leq \chi(G) \leq \lambda_{\max}(A) + 1.$$

This proves the left-hand inequality in (2.31).

Now choose an eigenvector $f \geq 0$ of A for $\lambda_{\max}(A)$, then $\dfrac{f}{\langle \mathbb{1}, f \rangle} \in \Lambda(G)$. By Motzkin-Straus theorem 2.8.1,

$$\frac{\lambda_{\max}(A) \, \|f\|^2}{|\langle \mathbb{1}, f \rangle|^2} = \left\langle \frac{f}{\langle \mathbb{1}, f \rangle}, A \frac{f}{\langle \mathbb{1}, f \rangle} \right\rangle \leq F(G) = \frac{\omega(G) - 1}{\omega(G)},$$

so

$$\lambda_{\max}(A) \, \|f\|^2 \leq \frac{\omega(G) - 1}{\omega(G)} |\langle \mathbb{1}, f \rangle|^2.$$

With the Cauchy-Schwarz inequality $|\langle \mathbb{1}, f \rangle|^2 \leq \|\mathbb{1}\|^2 \|f\|^2 \equiv |X| \, \|f\|^2$ the right-hand inequality in (2.31) follows. $\qquad\square$

Remark 2.8.5 Using $\omega(G) \leq \chi(G)$, inequality (2.32) also yields

$$\frac{|X|}{|X| - \lambda_{\max}(A)} \leq \chi(G),$$

which was already found by Cvetković [36].

The Motzkin-Straus theorem and formula (2.28) are very practical, as by substituting any $f \in \Lambda(G)$ into $\langle f, Af \rangle$ concrete estimates for $F(G)$ and thus also for $\omega(G)$ are obtained. This can lead to quite surprising statements. The following theorem was found in [84].

Theorem 2.8.6 *For every graph* $G = (X, E)$,

$$\omega(G) \geq \frac{|X|^2}{|X|^2 - 2|E|}. \tag{2.33}$$

Proof Let $n := |X|$ and $m := |E|$. We have

$$\langle \mathbb{1}, A\mathbb{1}\rangle = 2\sum_{e\in E} \mathbb{1}(\iota e)\mathbb{1}(\tau e) = 2\sum_{e\in E} 1 = 2|E| = 2m, \quad \frac{\mathbb{1}}{|X|} \in \Lambda(G),$$

and formula (2.28) states that

$$\frac{\omega(G)-1}{\omega(G)} \geq \left\langle \frac{\mathbb{1}}{|X|}, A\frac{\mathbb{1}}{|X|}\right\rangle = \frac{\langle \mathbb{1}, A\mathbb{1}\rangle}{|X|^2} \equiv \frac{2m}{n^2}.$$

It follows that

$$\omega(G) \geq \frac{n^2}{n^2 - 2m}.$$

$\square$

Example 2.8.7

(a) Let $n \geq 3$. The cycle C_n has n vertices and n edges,

$$\omega(C_n) = \begin{cases} 3, & n = 3, \\ 2, & n \geq 4, \end{cases}$$

$$\frac{|X|^2}{|X|^2 - 2|E|} = \frac{n^2}{n^2 - 2} = \frac{n}{n-2} \begin{cases} = 3, & n = 3, \\ \in (1, 2], & n \geq 4, \end{cases}$$

so the estimate from Theorem 2.8.6 is optimal for this case.

(b) The graph $K_{m,n}$ has $m + n$ vertices and mn edges, so

$$\frac{|X|^2}{|X|^2 - 2|E|} = \frac{(m+n)^2}{m^2 + n^2} = 1 + \frac{2mn}{m^2 + n^2} \in (1, 2],$$

and the lower estimate in (2.33) is again optimal, since $\omega(K_{m,n}) = 2$.

Remark 2.8.8 There are stronger advanced versions of Theorem 2.8.6. Namely, for all $m, n \in \mathbb{N}$ choose $r \in \{0, \ldots, m-1\}$ with $\frac{n-r}{m} \in \mathbb{Z}$ and define

$$t_m(n) := \frac{(m-1)(n^2 - r^2)}{2m} + \frac{r(r-1)}{2},$$

then the following statement holds (Turán's theorem): If a graph $G = (X, E)$ with n vertices does not contain a complete subgraph with m vertices, $\omega(G) \leq m - 1$, then G has at most $t_{m-1}(n)$ edges. For $|E| = t_{m-1}(n)$, G is isomorphic to the so-called Turán graph $T_{m-1}(n)$ (all Turán graphs are known). For different proof methods, we refer to [3, Chapter 36] and [45, Chapter 6].

It is interesting to compare Turán's theorem with our Theorem 2.8.6. From (2.33) it follows: for a graph $G = (X, E)$ with $|X| = n$ and $\omega(G) \leq m - 1$ we have

$$
m - 1 \geq \frac{n^2}{n^2 - 2|E|}, \quad \text{i.e. } |E| \leq \frac{n^2}{2}\left(1 - \frac{1}{m - 1}\right),
$$

and maximizing over E yields

$$
t_{m-1}(n) \leq \frac{n^2}{2}\left(1 - \frac{1}{m - 1}\right), \qquad \text{thus } t_m(n) \leq \frac{n^2}{2}\left(1 - \frac{1}{m}\right).
$$

One can show that the difference between the elementary upper bound and the exact value $t_m(n)$ remains "relatively small" for large n, see [84].

Exercise 2.8.1 From the article [88].

Let $G = (X, E)$ be a graph and $\overline{G}$ its complement graph. Let A and $\overline{A}$ be the adjacency matrices of G and $\overline{G}$. Denote $n := |X|$ and $m := |E|$. Show:

(a) $\lambda_{\max}(A) \leq \sqrt{2m\left(1 - \dfrac{1}{\omega(G)}\right)}$. Hint: For $f \in \mathbb{K}^X$ with $\|f\| = 1$ we have $|f|^2 \in \Lambda(G)$.

(b) $\lambda_{\max}(A) + \lambda_{\max}(\overline{A}) \leq \sqrt{\left(2 - \dfrac{1}{\omega(G)} - \dfrac{1}{\alpha(G)}\right)n(n-1)}$.

(c) $\lambda_{\max}(A) + \lambda_{\max}(\overline{A}) \leq \sqrt{\left(2 - \dfrac{1}{\chi(G)} - \dfrac{1}{\chi(\overline{G})}\right)n(n-1)}$.

2.9 Anticlique Number

In this section we prove some inequalities for the anticlique number $\alpha(G)$ (see Sect. 1.3). Some auxiliary constructions will also be actively used later. One of the first estimates comes from [38, Chapter 3.2].

Theorem 2.9.1 (Cvetković) *If A is the adjacency matrix of a graph G, then*

$$
\alpha(G) \leq \min \big\{ \text{the number of nonnegative eigenvalues of } A,
$$

$$
\text{the number of nonpositive eigenvalues of } A \big\}.
$$

Proof Denote $n := |X|$ and $m := \alpha(G)$ and let $Y \subset X$ be an anticlique with $|Y| = m$. Then $A_{x,y} = 0$ for all $x, y \in Y$, i.e., the adjacency matrix A' of the induced graph G_Y is the zero matrix. Since A' is obtained by deleting $n - m$ columns and $n - m$ rows in A, the interlacing principle for matrices (Corollary 2.1.9) gives

$$\lambda_k(A) \le \lambda_k(A') \le \lambda_{k+n-m}(A) \text{ for all } k \in \{1, \dots, m\}.$$

Since $\lambda_k(A') = 0$ for all k, we get $\lambda_m(A) \le 0 \le \lambda_{n-m+1}(A)$, and it follows

$$m \le \max_{\lambda_k(A) \le 0} k = \text{the number of nonpositive eigenvalues of } A,$$

$$m \le \max_{\lambda_{n-k+1}(A) \ge 0} k = \text{the number of nonnegative eigenvalues of } A.$$

$\square$

Example 2.9.2

(a) The adjacency matrix of K_n has only one nonnegative eigenvalue $n - 1$ (Example 1.6.2), so we get the inequality $\alpha(K_n) \le 1$. This is optimal: Any anticlique in K_n consist of a single vertex.
(b) The adjacency matrix of $K_{m,n}$ has $n + m - 1$ nonnegative and $n + m - 1$ nonpositive eigenvalues (Example 1.6.7), so we get $\alpha(K_{m,n}) \le n + m - 1$, where the actual value is $\alpha(K_{m,n}) = \max\{m, n\}$ (choose the largest partition component).
(c) For the cycle C_n, we clearly have $\alpha(C_n) = [\frac{n}{2}]$, where the number of nonnegative or non-positive eigenvalues of A is given by

$$\#\left\{ k \in \{0, \dots, n - 1\} : 2\cos\frac{2\pi k}{n} \ge 0 \right\} \equiv \left[\frac{n}{2}\right] + 1$$

$$\text{or } \#\left\{ k \in \{0, \dots, n - 1\} : 2\cos\frac{2\pi k}{n} \le 0 \right\} \equiv \left[\frac{n}{2}\right] + 1.$$

Now we will also use the Laplace matrix for the investigation of anticliques. The approach here is much more direct than with A and mostly follows [98]. The following definitions and identities will also be actively used in subsequent chapters.

Definition 2.9.3 Let $G = (X, E)$ be a graph. For $Y \subset X$, define the **boundary ∂Y of Y in G** by

$$\partial Y := \left\{ e \in E : e \text{ connects a vertex in } Y \text{ with a vertex in } X \setminus Y \right\}.$$

Note that $Y \subset X$ but $\partial Y \subset E$. Obviously, $\partial Y = \partial(X \setminus Y)$.

The next identity explains the special significance of the Laplace matrix in the investigation of induced subgraphs.

Lemma 2.9.4 (Indicator Functions and L) *Let L be the Laplace matrix of a graph $G = (X, E)$ and $Y \subset X$, then*

$$\langle \mathbb{1}_Y, L\mathbb{1}_Y \rangle = |\partial Y|.$$

Proof Choose an arbitrary orientation on G, then (Lemma 1.5.5)

$$\langle \mathbb{1}_Y, L\mathbb{1}_Y \rangle = \sum_{e \in E} \left| \mathbb{1}_Y(\iota e) - \mathbb{1}_Y(\tau e) \right|^2,$$

$$= \sum_{\substack{e \text{ from } Y \text{ to } Y}} \big| \underbrace{\mathbb{1}_Y(\iota e) - \mathbb{1}_Y(\tau e)}_{=1-1=0} \big|^2 + \sum_{\substack{e \text{ from } X\backslash Y \text{ to } X\backslash Y}} \big| \underbrace{\mathbb{1}_Y(\iota e) - \mathbb{1}_Y(\tau e)}_{=0-0=0} \big|^2$$

$$+ \sum_{\substack{e \text{ from } Y \text{ to } X\backslash Y}} \big| \underbrace{\mathbb{1}_Y(\iota e) - \mathbb{1}_Y(\tau e)}_{=1-0 \text{ or } 0\text{-}1} \big|^2$$

$$= \sum_{\substack{e \text{ from } Y \text{ to } X\backslash Y}} 1 \equiv \sum_{e \in \partial Y} 1 = |\partial Y|.$$

$\square$

We will also need the following simple estimate.

Lemma 2.9.5 *Let $Y \subset X$ and $s \in \mathbb{R}$, then $\|\mathbb{1}_Y - s\mathbb{1}\|^2$ is minimal for $s = \frac{|Y|}{|X|}$,*

$$\|\mathbb{1}_Y - s\mathbb{1}\|^2 = s(1-s)|X| \text{ for } s := \frac{|Y|}{|X|}.$$

Proof It holds

$$\|\mathbb{1}_Y - s\mathbb{1}\|^2 = \langle \mathbb{1}_Y - s\mathbb{1}, \mathbb{1}_Y - s\mathbb{1} \rangle$$

$$= \langle \mathbb{1}_Y, \mathbb{1}_Y \rangle - 2s\langle \mathbb{1}_Y, \mathbb{1} \rangle + s^2 \langle \mathbb{1}, \mathbb{1} \rangle = |Y| - 2s|Y| + s^2|X|.$$

The minimum of the quadratic polynomial is reached for $s = \frac{|Y|}{|X|}$, and for this value of s it holds

$$\|\mathbb{1}_Y - s\mathbb{1}\|^2 = s|X| - 2s^2|X| + s^2|X| = s(1-s)|X|.$$

$\square$

The next statements follow the article by Godsil and Newman [55].

Theorem 2.9.6 *Let $G = (X, E)$ be a graph with at least one edge, L its Laplace matrix and $Y \subset X$ an anticlique, then*

$$|Y| \leq |X|\left(1 - \frac{\deg_{\mathrm{avg}} Y}{\lambda_{\max}(L)}\right) \text{ with } \deg_{\mathrm{avg}} Y := \frac{\sum\limits_{y \in Y} \deg y}{|Y|}. \tag{2.34}$$

Proof Thanks to the min-max formula it holds that

$$\lambda_{\max}(L) \geq \frac{\langle f, Lf \rangle}{\|f\|^2} \text{ for all } f \in \mathbb{K}^X, \ f \not\equiv 0.$$

We will insert $f := \mathbb{1}_Y - s\mathbb{1}$ with $s \in \mathbb{R}$. Since $L\mathbb{1} = 0$, we have $Lf = L\mathbb{1}_Y$, and with Lemma 2.9.4 we see that the number

$$\langle f, Lf \rangle = \langle f, L\mathbb{1}_Y \rangle = \langle Lf, \mathbb{1}_Y \rangle = \langle L\mathbb{1}_Y, \mathbb{1}_Y \rangle = |\partial Y|$$

is independent of s. So it is advantageous to minimize the norm of f to have a possibly best lower estimate for $\lambda_{\max}(L)$. According to Lemma 2.9.5, $s := \frac{|Y|}{|X|}$ must then be chosen, and one obtains

$$\lambda_{\max}(L) \geq \frac{|\partial Y|}{s(1-s)|X|}. \tag{2.35}$$

Now we use the fact that Y is an anticlique. In this case, every edge that has one of the end vertices in Y is in $|\partial Y|$. Since each vertex $y \in Y$ is an end vertex for exactly $\deg y$ edges and these edges are all distinct (there are no edges within Y), it follows that

$$|\partial Y| = \sum_{y \in Y} \deg y = (\deg_{\mathrm{avg}} Y)|Y| \equiv (\deg_{\mathrm{avg}} Y)s|X|.$$

By substituting into (2.35) we get

$$\lambda_{\max}(L) \geq \frac{(\deg_{\mathrm{avg}} Y)s|X|}{s(1-s)|X|} \equiv \frac{\deg_{\mathrm{avg}} Y}{1-s},$$

and it follows

$$\frac{|Y|}{|X|} \equiv s \leq 1 - \frac{\deg_{\mathrm{avg}} Y}{\lambda_{\max}(L)}.$$

$\square$

The inequality (2.34) does not yet provide an estimate for the anticlique number, as the right-hand side still depends on Y. However, one can estimate $\deg_{\mathrm{avg}} Y$ by global invariants.

Corollary 2.9.7 (Anticlique Number and L-eigenvalues) *Let $G = (X, E)$ be a graph with at least one edge, L its Laplace matrix, then*

$$\alpha(G) \le |X|\left(1 - \frac{\deg_{\min} G}{\lambda_{\max}(L)}\right), \quad \deg_{\min} G := \min_{x \in X} \deg x.$$

Proof Since $\deg_{\mathrm{avg}} Y \ge \deg_{\min} G$ for all $Y \subset X$, the assertion follows directly from theorem 2.9.6. $\qquad\qquad\qquad\qquad\qquad\qquad\qquad\qquad\qquad\qquad\qquad\square$

Remark 2.9.8 If G is a k-regular graph, then $\lambda_{\max}(L) = k - \lambda_{\min}(A)$ and $\deg_{\min} G = k$, so

$$\alpha(G) \le |X|\frac{-\lambda_{\min}(A)}{k - \lambda_{\min}(A)}, \quad \text{if } G \text{ is } k\text{-regular.}$$

Using Proposition 1.3.4 it follows

$$\chi(G) \ge \frac{|X|}{\alpha(G)} \ge \frac{k - \lambda_{\min}(A)}{-\lambda_{\min}(A)} \equiv 1 + \frac{\lambda_{\max}(A)}{-\lambda_{\min}(A)},$$

since in this case $\lambda_{\max}(A) = k$ (see Exercise 1.6.6). This provides an alternative proof for the Hoffman estimate (Theorem 2.6.4) for regular graphs.

Remark 2.9.9 Further estimates of graph invariants by the eigenvalues of the graph matrices are discussed in many books [11, 24, 38, 39, 41, 100, 103].

Exercise 2.9.1 From the article [61]
   Let $G = (X, E)$ be a graph with at least one edge and let A be its adjacency matrix. We want to prove the inequality

$$\alpha(G) \le |X| \frac{-\lambda_{\min}(A)\,\lambda_{\max}(A)}{(\deg_{\min} G)^2 - \lambda_{\min}(A)\,\lambda_{\max}(A)} \tag{2.36}$$

Let $n := |X|$ and $k := \alpha(G)$ and let $Y \subset X$ be an anticlique with $|Y| = k$. W.l.o.g, we assume that the vertices of Y are the first k vertices of X, i.e.

$$A = \begin{pmatrix} 0 & S \\ S^* & T \end{pmatrix}$$

with a $k \times (n-k)$ block S and a $(n-k) \times (n-k)$ block T. Furthermore, we consider the $n \times 2$ matrix

$$\Theta := \begin{pmatrix} \dfrac{\mathbb{1}_k}{\sqrt{k}} & 0 \\ 0 & \dfrac{\mathbb{1}_{n-k}}{\sqrt{n-k}} \end{pmatrix} \quad \text{with } \mathbb{1}_m := \left.\begin{pmatrix} 1 \\ \cdot \\ \cdot \\ \cdot \\ 1 \end{pmatrix}\right\} m \text{ rows}$$

as well as the 2×2 matrix $B := \Theta^* A \Theta$. Show:

(a) $\Theta^* \Theta = I_2$,

(b) $B_{1,1} = 0$, $B_{1,2} = B_{2,1} = \dfrac{\displaystyle\sum_{y \in Y} \deg Y}{\sqrt{k(n-k)}}$ and $B_{2,2} \geq 0$,

(c) $\lambda_2(B) \geq 0$,

(c) $-\lambda_{\min}(A)\,\lambda_{\max}(A) \geq -\lambda_1(B)\lambda_2(B) = -\det B$,

(d) $-\det B \geq \dfrac{k}{n-k}\,(\deg_{\min} G)^2$,

and prove the desired inequality (2.36).

Partitions and Eigenfunctions

3

Can one divide a graph into several connected parts with "good" properties? Possible mathematical formulations for this general question will be discussed in this chapter. In particular, we will see how the signs and the zeros of the eigenfunctions of graph matrices can be used for these purposes.

3.1 Cheeger Inequality

Let $G = (X, E)$ be a <u>connected</u> graph. We deal with the following informally posed question: How can one divide X into two approximately equally large parts Y and $X \setminus Y$ (with $Y \subsetneq X$), so that the boundary ∂Y is the "smallest"? For the "size" of $\partial Y \subset E$ there is a natural measure:

$$|\partial Y| \equiv \text{the number of edges in } \partial Y.$$

The "size" of $Y \subset X$ can be characterized by (at least) two numbers:

$$|Y| \equiv \text{the number of vertices in } Y,$$

$$\deg Y := \sum_{y \in Y} \deg y \equiv 2 \times \text{the number of edges in } Y \\ + \text{ the number of edges in } |\partial Y|.$$

So we define for each subset $Y \subset X$ with $Y \notin \{\varnothing, X\}$

$$\theta'(Y) := \frac{|\partial Y|}{\min\{|Y|, |X \setminus Y|\}} \equiv \max\left\{\frac{|\partial Y|}{|Y|}, \frac{|\partial(X \setminus Y)|}{|X \setminus Y|}\right\},$$

$$\theta(Y) := \frac{|\partial Y|}{\min\{\deg Y, \deg X \setminus Y\}} \equiv \max\left\{\frac{|\partial Y|}{\deg Y}, \frac{|\partial(X \setminus Y)|}{\deg X \setminus Y}\right\}.$$

© The Author(s), under exclusive license to Springer Nature Switzerland AG 2025
K. Naderi, K. Pankrashkin, *Introduction to Spectral Graph Theory*, Compact
Textbooks in Mathematics, https://doi.org/10.1007/978-3-032-01708-6_3

The numbers

$$h'(G) := \min_{\varnothing \neq Y \subsetneq X} \theta'(Y), \quad h(G) := \min_{\varnothing \neq Y \subsetneq X} \theta(Y)$$

are called **isoperimetric** or **Cheeger constants** of G with respect to vertex counting or edge counting, and they can be interpreted as numerical indicators of the "minimal density" in the graph. One is interested in the subsets $Y \subset X$, so that $\theta'(Y)$ or $\theta(Y)$ are as close as possible to the optimal values $h'(G)$ or $h(G)$.

In practice, $h(G)$ is often used, as $\deg Y$ is seen as a "better" measure of Y. But we will first consider the constant $h'(G)$, as it is closely related to the well-known Laplace matrix.

Theorem 3.1.1 *Let $G = (X, E)$ be a connected graph with at least two vertices and L its Laplace matrix, then:*

(a) For each $\varnothing \neq Y \subsetneq X$ it holds that

$$\theta'(Y) \geq \max\{s, 1 - s\} \lambda_2(L) \text{ with } s := \frac{|Y|}{|X|},$$

$$\frac{|X| |\partial Y|}{|Y| |X \setminus Y|} \geq \lambda_2(L), \tag{3.1}$$

(b) and

$$h'(G) \geq \frac{\lambda_2(L)}{2}. \tag{3.2}$$

Proof

(a) We have

$$\lambda_2(L) \leq \frac{\langle f, Lf \rangle}{\|f\|^2} \text{ for each } f \in \mathbb{K}^X \text{ with } f \not\equiv 0 \text{ and } \langle \mathbb{1}, f \rangle = 0,$$

see Corollary 2.1.6. We want to insert $f := \mathbb{1}_Y - s\mathbb{1}$ with suitable $s \in \mathbb{R}$. The condition $\langle \mathbb{1}, f \rangle = 0$ has the form $0 = \langle \mathbb{1}, \mathbb{1}_Y - s\mathbb{1} \rangle = |Y| - s|X|$, so $s := |Y|/|X|$ is the only possible value. Thanks to Lemmas 2.9.4 and 2.9.5 we get

$$\|f\|^2 = s(1 - s)|X| \equiv (1 - s)|Y|, \qquad \langle f, Lf \rangle = \langle \mathbb{1}_Y, L\mathbb{1}_Y \rangle = |\partial Y|,$$

so

$$\lambda_2(L) \leq \frac{|\partial Y|}{(1 - s)|Y|}, \quad \text{that is } \frac{|\partial Y|}{|Y|} \geq (1 - s)\lambda_2(L), \quad s := \frac{|Y|}{|X|}.$$

If we insert $X \setminus Y$ instead of Y, we get

$$\frac{|\partial(X \setminus Y)|}{|X \setminus Y|} \geq s\lambda_2(L), \tag{3.3}$$

so overall $\theta'(Y) \geq \max\{1 - s, s\}\lambda_2(L)$. By inserting the explicit value of s into (3.3), we get (3.1).

(b) In (a) the subset Y was arbitrary, for any $s \in (0, 1)$ one has $\max\{1 - s, s\} \geq \frac{1}{2}$, and (3.2) follows.

$\square$

Now we discuss the values $\theta(Y)$ and $h(G)$. We note that in the definition of $\theta(Y)$ the expression

$$\frac{|\partial Y|}{\deg Y} \equiv \frac{\langle \mathbb{1}_Y, L\mathbb{1}_Y \rangle}{\langle \mathbb{1}_Y, D\mathbb{1}_Y \rangle}$$

appears. So we want to understand whether the quotients

$$\frac{\langle f, Lf \rangle}{\langle f, Df \rangle} \text{ with } 0 \neq f \in \mathbb{K}^X$$

are related to the eigenvalues of any matrix.

For $a \in \mathbb{R}$ consider the diagonal matrix D^a,

$$(D^a)_{x,y} = (\deg x)^a \delta_{x,y}, \quad x, y \in X.$$

Obviously $D^1 = D$, $D^0 = I$ and $D^a D^b = D^{a+b}$ for all $a, b \in \mathbb{R}$. Furthermore, we consider the vectors

$$d^a := D^a \mathbb{1} : X \ni x \mapsto (\deg x)^a, \quad d := d^1, \quad d^0 := \mathbb{1},$$

then we also have $D^a d^b = d^{a+b}$ for all $a, b \in \mathbb{R}$.

Since G is connected, $D^a : \mathbb{K}^X \to \mathbb{K}^X$ is an isomorphism for every $a \in \mathbb{R}$. For all $f \in \mathbb{K}^X$ and $g := D^{\frac{1}{2}} f$ we have

$$\frac{\langle f, Lf \rangle}{\langle f, Df \rangle} = \frac{\langle D^{-\frac{1}{2}}g, LD^{-\frac{1}{2}}g \rangle}{\langle D^{-\frac{1}{2}}g, DD^{-\frac{1}{2}}g \rangle} = \frac{\langle g, D^{-\frac{1}{2}}LD^{-\frac{1}{2}}g \rangle}{\langle g, g \rangle},$$

and this is exactly the Rayleigh quotient of g for

$$N := D^{-\frac{1}{2}}LD^{-\frac{1}{2}} \quad \textbf{(the normalized Laplace matrix for } G\textbf{)}.$$

The matrix N is only defined when G has no isolated vertices (for the connected graphs with at least two vertices this assumption is always fulfilled).

Theorem 3.1.2 *Let $G = (X, E)$ be a connected graph with at least two vertices and N its normalized Laplace matrix. Then N defines a self-adjoint linear mapping in $\mathbb{K}^X$ with $\lambda_1(N) = 0$ and $\ker N = \mathbb{K}d^{\frac{1}{2}}$, and*

$$0 < \lambda_2(N) = \min_{\substack{0 \neq f \in \mathbb{K}^X \\ f \perp d}} \frac{\langle f, Lf \rangle}{\langle f, Df \rangle}. \tag{3.4}$$

Proof The self-adjointness follows from

$$N^* = \left(D^{-\frac{1}{2}}LD^{-\frac{1}{2}}\right)^* = (D^{-\frac{1}{2}})^* L^* (D^{-\frac{1}{2}})^* = D^{-\frac{1}{2}}LD^{-\frac{1}{2}} = N.$$

For every $g \in \mathbb{K}^X$ we have

$$\langle g, Ng \rangle = \langle g, D^{-\frac{1}{2}}LD^{-\frac{1}{2}}g \rangle = \langle D^{-\frac{1}{2}}g, LD^{-\frac{1}{2}}g \rangle \geq 0,$$

so all eigenvalues of N are nonnegative by the min-max principle.

Since $D^{-\frac{1}{2}} : \mathbb{K}^X \to \mathbb{K}^X$ is bijective, it follows

$$\ker N = \left\{ g \in \mathbb{K}^X : D^{-\frac{1}{2}}LD^{-\frac{1}{2}}g = 0 \right\}$$

$$= \left\{ g \in \mathbb{K}^X : D^{-\frac{1}{2}}g \in \ker L \equiv \mathbb{K}\mathbb{1} \right\} = \mathbb{K}d^{\frac{1}{2}},$$

so the first eigenvalue 0 is simple and thus the second eigenvalue is strictly positive. By the min-max formula (Corollary 2.1.6) we have

$$\lambda_2(N) = \min_{\substack{0 \neq g \in \mathbb{K}^X \\ \langle d^{\frac{1}{2}}, g \rangle = 0}} \frac{\langle g, Ng \rangle}{\langle g, g \rangle} = \left(\text{write } g = D^{\frac{1}{2}}f \text{ with } f \in \mathbb{K}^X \right)$$

$$= \min_{\substack{0 \neq f \in \mathbb{K}^X \\ \langle d^{\frac{1}{2}}, D^{\frac{1}{2}}f \rangle = 0}} \frac{\langle D^{\frac{1}{2}}f, (D^{-\frac{1}{2}}LD^{-\frac{1}{2}})D^{\frac{1}{2}}f \rangle}{\langle D^{\frac{1}{2}}f, D^{\frac{1}{2}}f \rangle} \equiv \min_{\substack{0 \neq f \in \mathbb{K}^X \\ \langle d, f \rangle = 0}} \frac{\langle f, Lf \rangle}{\langle f, Df \rangle}.$$

$$\square$$

For further properties of the eigenvalues of N see Exercise 3.1.3.

Theorem 3.1.3 *Let $G = (X, E)$ be a connected graph with normalized Laplace matrix N, then:*

(a) For every $\varnothing \neq Y \subsetneq X$ it holds

$$\theta(Y) \geq \max\{s, 1 - s\}\,\lambda_2(N) \text{ with } s := \frac{\deg Y}{\deg X},$$

$$\frac{\deg X\,|\partial Y|}{\deg Y(\deg X - \deg Y)} \geq \lambda_2(N), \tag{3.5}$$

(b) in particular

$$h(G) \geq \frac{\lambda_2(N)}{2}.$$

Proof The proof runs very similar to what was done in Theorem 3.1.1.

(a) We want to insert a function $f = \mathbb{1}_Y - s\mathbb{1}$ into the right-hand side of (3.4). From the condition

$$0 = \langle d, f \rangle \equiv \langle d, \mathbb{1}_Y \rangle - s\langle d, \mathbb{1} \rangle = \deg Y - s \deg X$$

one obtains $s = \deg Y / \deg X$. We have $\langle f, Lf \rangle = \langle \mathbb{1}_Y, L\mathbb{1}_Y \rangle = |\partial Y|$ and

$$\langle f, Df \rangle = \langle \mathbb{1}_Y - s\mathbb{1}, D(\mathbb{1}_Y - s\mathbb{1}) \rangle = \langle \mathbb{1}_Y, D\mathbb{1}_Y \rangle - 2s\langle \mathbb{1}, D\mathbb{1}_Y \rangle + s^2\langle \mathbb{1}, D\mathbb{1} \rangle$$

$$= \deg Y - 2s \deg Y + s^2 \deg X = s \deg X - 2s^2 \deg X + s^2 \deg X$$

$$= (1 - s)s \deg X = (1 - s) \deg Y,$$

thus

$$\lambda_2(N) \leq \frac{\langle f, Lf \rangle}{\langle f, Df \rangle} = \frac{|\partial Y|}{(1 - s) \deg Y}, \quad \text{i.e.} \quad \frac{|\partial Y|}{\deg Y} \geq (1 - s)\lambda_2(N),$$

which is equivalent to (3.5). If one substitutes $X \setminus Y$ for Y, one obtains

$$\frac{|\partial(X \setminus Y)|}{\deg(X \setminus Y)} \geq s\lambda_2(N),$$

from which follows $\theta(Y) \geq \max\{1 - s, s\}\lambda_2(N)$.

(b) In the above estimate, $s \in (0, 1)$ holds, thus $\max\{1 - s, s\} \geq \frac{1}{2}$ and

$$\theta(Y) \geq \frac{\lambda_2(N)}{2} \text{ for all } \varnothing \neq Y \subset X.$$

$\square$

It is interesting that $h(G)$ can be estimated <u>from both sides</u> with the help of $\lambda_2(N)$.

Theorem 3.1.4 (Cheeger Inequality for Edge Counting) *Let $G = (X, E)$ be a connected graph and N its normalized Laplace matrix, then*

$$\frac{\lambda_2(N)}{2} \leq h(G) \leq \sqrt{2\lambda_2(N)}$$

or, equivalently,

$$\frac{h(G)^2}{2} \leq \lambda_2(N) \leq 2h(G).$$

As a preparation, we prove:

Lemma 3.1.5 *Let $G = (X, E)$ be a graph with an orientation and $\vec{\nabla}$ its oriented edge-vertex matrix,*

$$\vec{\nabla} : \mathbb{K}^X \to \mathbb{K}^E, \quad (\vec{\nabla} f)(e) = f(\iota e) - f(\tau e) \text{ for all } f \in \mathbb{K}^X \text{ and } e \in E.$$

Then for all $f \in \mathbb{R}^X$ one has the identity

$$\sum_{e \in E} |\vec{\nabla} f(e)| = \int_{-\infty}^{\infty} \left| \partial\{x \in X : f(x) > t\} \right| \, dt.$$

Proof Let $f \in \mathbb{R}^X$. For $e \in E$ denote

$$a_e := \min\left\{ f(\iota e), f(\tau e) \right\}, \quad b_e := \max\left\{ f(\iota e), f(\tau e) \right\}, \quad I_e := [a_e, b_e),$$

then

$$\sum_{e \in E} |\vec{\nabla} f(e)| = \sum_{e \in E} |I_e|. \tag{3.6}$$

For arbitrary $e \in E$ and $t \in \mathbb{R}$ the condition $e \in \partial\{x \in X : f(x) > t\}$ is equivalent to $t \in I_e$, thus

$$\left| \partial\{x \in X : f(x) > t\} \right| = \left| \partial\{e \in E : t \in I_e\} \right| = \sum_{e \in E} \mathbb{1}_{I_e}(t),$$

$$\int_{-\infty}^{\infty} \left| \partial\{x \in X : f(x) > t\} \right| \, dt = \int_{-\infty}^{\infty} \sum_{e \in E} \mathbb{1}_{I_e}(t) \, dt = \sum_{e \in E} \int_{-\infty}^{\infty} \mathbb{1}_{I_e}(t) \, dt$$

$$= \sum_{e \in E} |I_e| \stackrel{(3.6)}{=} \sum_{e \in E} |\vec{\nabla} f(e)|.$$

$\square$

Proof of Theorem 3.1.4 (Cheeger Inequality) We write $\lambda_k := \lambda_k(N)$. The lower estimate for $h(G)$ is already proven (Theorem 3.1.3). We show

$$\frac{h(G)^2}{2} \le \lambda_2,$$

from which the rest follows by elementary rearrangements.

Let f' be an eigenvector of N for λ_2. Since the matrix and the eigenvalue are real, we can assume w.l.o.g. that $f' \in \mathbb{R}^X$ (otherwise we take $\operatorname{Re} f'$ or $\operatorname{Im} f'$ instead of f'). We have $\lambda_1 < \lambda_2$, thus $\langle d^{\frac{1}{2}}, f' \rangle = 0$. Since $d^{\frac{1}{2}} > 0$, f' takes negative <u>and</u> positive values, the same holds for $f := D^{-\frac{1}{2}} f'$. We have

$$Lf = D^{\frac{1}{2}} N D^{\frac{1}{2}} f = D^{\frac{1}{2}} N f' = D^{\frac{1}{2}} \lambda_2 f' = D^{\frac{1}{2}} \lambda_2 D^{\frac{1}{2}} f = \lambda_2 D f. \qquad (3.7)$$

Consider the disjoint non-empty sets

$$X_+ := \{ x \in X : f(x) > 0 \}, \quad X_- := \{ x \in X : f(x) < 0 \}.$$

W.l.o.g., we assume that $\deg X_+ \le \deg X_-$ (otherwise, we take $-f'$ instead of f' and $-f$ instead of f from the beginning). Consider

$$g := \frac{f + |f|}{2}, \quad \text{i.e. } g : X \ni x \mapsto \begin{cases} f(x), & x \in X_+, \\ 0, & \text{otherwise,} \end{cases}$$

then we obviously have $g \ge 0$ with $f(x) \le g(x)$ for all $x \in X$.

For $u, v \in \mathbb{K}^X$ we write

$$\langle u, v \rangle_+ := \sum_{x \in X_+} \overline{u(x)} v(x).$$

We now want to estimate the number

$$\lambda_2 \equiv \lambda_2 \frac{\langle f, Df \rangle_+}{\langle f, Df \rangle_+} \equiv \frac{\langle f, \lambda_2 Df \rangle_+}{\langle f, Df \rangle_+} \stackrel{(3.7)}{=} \frac{\langle f, Lf \rangle_+}{\langle f, Df \rangle_+}$$

from below. We have $\langle f, Df \rangle_+ = \langle g, Dg \rangle_+ = \langle g, Dg \rangle$ and

$$\langle f, Lf \rangle_+ = \sum_{\substack{x \in X_+ \\ =g(x)>0}} \underbrace{f(x)}_{=g(x)>0} \sum_{y \sim x} \Big(\underbrace{f(x)}_{} - \underbrace{f(y)}_{\le g(y)} \Big)$$

$$\geq \sum_{x\in X_+} g(x) \sum_{y\sim x} \big(g(x) - g(y)\big) = \sum_{x\in X} g(x) \sum_{y\sim x} \big(g(x) - g(y)\big)$$

$$(\text{since } g(x) = 0 \text{ for all } x \in X \setminus X_+)$$

$$= \langle g, Lg \rangle = \sum_{e\in E} \big|g(\iota e) - g(\tau e)\big|^2 \equiv \sum_{e\in E} \big|\vec{\nabla} g(e)\big|^2.$$

Therefore,

$$\lambda_2 \geq \frac{\sum_{e\in E} \big|\vec{\nabla} g(e)\big|^2}{\langle g, Dg \rangle}. \tag{3.8}$$

Now consider the function $g^2 : X \ni x \mapsto g(x)^2$, then

$$\left(\sum_{e\in E} \big|\vec{\nabla}(g^2)(e)\big|\right)^2 = \left(\sum_{e\in E} \big|g(\iota e)^2 - g(\tau e)^2\big|\right)^2$$

$$= \left(\sum_{e\in E} \big|g(\iota e) - g(\tau e)\big|\,\big|g(\iota e) + g(\tau e)\big|\right)^2 \tag{3.9}$$

$$(\text{Cauchy-Schwarz}) \quad \leq \sum_{e\in E} \big|g(\iota e) - g(\tau e)\big|^2 \sum_{e\in E} \big|g(\iota e) + g(\tau e)\big|^2.$$

The last factor can be estimated from above using $(a + b)^2 \leq 2(a^2 + b^2)$ for all $a, b \in \mathbb{R}$:

$$\sum_{e\in E} \big|g(\iota e) + g(\tau e)\big|^2 \leq 2 \sum_{e\in E} \left(\big|g(\iota e)\big|^2 + \big|g(\tau e)\big|^2\right) = 2$$

$$\sum_{x\in X} (\deg x)\big|g(x)\big|^2 \equiv 2\langle g, Dg \rangle,$$

and from (3.9) it follows

$$\left(\sum_{e\in E} \big|\vec{\nabla}(g^2)(e)\big|\right)^2 \leq 2 \sum_{e\in E} \big|\vec{\nabla} g(e)\big|^2 \langle g, Dg \rangle.$$

By substituting into (3.8) we obtain

$$\lambda_2 \geq \frac{\left(\sum_{e\in E} \big|\vec{\nabla}(g^2)(e)\big|\right)^2}{2\langle g, Dg \rangle^2}. \tag{3.10}$$

For $t \in \mathbb{R}$ we consider the sets

$$X_t := \{ x \in X : g(x)^2 > t \}.$$

For all $t < 0$ one has $X_t = X$ and thus $\partial X_t = \varnothing$. For $t \geq 0$, $X_t \subset X_+$ and thus

$$\deg X_t \leq \deg X_+ \leq \deg X_- \leq \deg X \setminus X_+ \leq \deg X \setminus X_t,$$

so

$$\theta(X_t) \equiv \frac{|\partial X_t|}{\min \left\{ \deg X_t, \deg X \setminus X_t \right\}} = \frac{|\partial X_t|}{\deg X_t}.$$

Denote

$$\theta_g := \min_{t \geq 0} \theta(X_t),$$

then $|\partial X_t| \geq \theta_g \deg X_t$ for all $t \geq 0$. With the help of Lemma 3.1.5 we obtain

$$\sum_{e \in E} |\vec{\nabla}(g^2)(e)| = \int_{-\infty}^{\infty} |\partial X_t| \, dt = \int_0^{\infty} |\partial X_t| \, dt \geq \theta_g \int_0^{\infty} \deg X_t \, dt$$

$$= \theta_g \int_0^{\infty} \sum_{x \in X} \underbrace{\mathbb{1}_{X_t}(x)}_{=\mathbb{1}_{[0,g(x)^2)}(t)} \deg x \, dt$$

$$= \theta_g \sum_{x \in X} \deg x \int_0^{\infty} \mathbb{1}_{[0,g(x)^2)}(t) \, dt = \theta_g \sum_{x \in X} (\deg x) g(x)^2 \equiv \theta_g \langle g, Dg \rangle.$$

From (3.10) it follows

$$\lambda_2 \geq \frac{(\theta_g \langle g, Dg \rangle)^2}{2 \langle g, Dg \rangle^2} = \frac{\theta_g^2}{2}.$$

Since obviously $\theta_g \geq h(G)$ holds, the claim follows. $\qquad \square$

Remark 3.1.6 As an exercise, one can similarly estimate $h'(G)$. However, one obtains the much weaker inequality

$$\lambda_2(L) \geq \frac{h'(G)^2}{2 \deg_{\max} G},$$

which was found by Dodziuk [46]. Later, Mohar [82] proved the better estimate

$$h'(G) \leq \sqrt{\lambda_2(L)\big(2 \deg_{\max} G - \lambda_2(L)\big)}$$

Other versions of the Cheeger inequalities and their applications are discussed in the book by Chung [27].

Remark 3.1.7 The Cheeger constant was originally introduced in the context of Riemannian manifolds. Let M be a compact n-dimensional Riemannian manifold, then the Cheeger constant $h(M)$ of M is defined as

$$h(M) := \inf_{S} \frac{\mathrm{vol}_{n-1} S}{\min\{\mathrm{vol}_n M_1, \mathrm{vol}_n M_2\}},$$

where the infimum is calculated over all compact hypersurfaces $S \subset M$ that divide M into two parts M_1 and M_2. Cheeger [26] showed that $h(M)$ can be estimated by the first nontrivial eigenvalue of the Laplace-Beltrami operator on M (similar results also exist for the case when M is a manifold with boundary).

Cheeger's idea was later transferred to partitions of graphs: Initial estimates and their importance for graph theory were recognized by Alon and Milman [4, 6]. Theorem 3.1.4 was proved by Dodziuk and Kendall [47], the proof presented here follows the lecture notes by Post [93]. Grieser [59] showed that some constructions of graph theory around the Cheeger inequality can also be brought back into the continuous world to gain important new information about the eigenvalues of Laplace-Beltrami operators.

Remark 3.1.8 In the proof of Theorem 3.1.4, we see that a suitable set Y is constructed in the form $Y = X_t$. In practice, the set $Y := X_+$ often yields good results. The partitioning of graphs using the eigenfunctions of graph matrices is the main tool in spectral cluster analysis, which plays an important role in data science. Higher eigenvalues and eigenfunctions can also be included when looking for partitions into more than two parts: For a good overview of the important questions and methods, we refer to the article by von Luxburg [110], the lecture notes by Mahoney [79] and Trevisan [105], as well as the original article by Lee et al. [75]. The partitions using the sign of the eigenfunctions have the advantage that in most cases a controllable number of connected subgraphs are created: The next sections are dedicated to the corresponding results.

Remark 3.1.9 In recent years, Cheeger inequalities on hypergraphs have also been actively investigated, see e.g. the review article by Mulas et al. [86].

Exercise 3.1.1 Let G be a graph with Laplace matrix L. Adapt the proof of Theorem 3.1.4 to prove the inequality

$$\lambda_2(L) \geq \frac{h'(G)^2}{2 \deg_{\max} G}.$$

Exercise 3.1.2 From the article [90].

Let $G = (X, E)$ be a connected graph.

1. Let $K \subset X$ be a separating vertex set and $G_i = (X_i, E_i)$ a connected component of $G \setminus K$. Show that for all i, $|\partial X_i| \leq |X_i| \cdot |K|$.

2. Derive: $h'(G) \le \kappa(G)$.
3. Let $\varnothing \ne Y \subsetneq X$. Show: $|\partial Y| \ge \kappa'(G)$.
4. Prove the inequalities

$$\frac{2\kappa(G)}{|X|} \le \frac{2\kappa'(G)}{|X|} \le h'(G) \le \kappa(G) \le \kappa'(G).$$

Exercise 3.1.3 Let $G = (X, E)$ be a graph without isolated vertices and N its normalized Laplace matrix. Let $n := |X|$. Show:

1. $\operatorname{tr} N = n$.
2. $\lambda_2(N) \le \frac{n}{n-1} \le \lambda_n(N)$.
3. $\lambda_2(N) > 1$ if and only if G is complete. In this case, $\lambda_2(N) = \frac{n}{n-1}$.
4. $\lambda_n(N) \le 2$.
5. $\lambda_n(N) = 2$ if and only if G has a bipartite connected component.
6. G is bipartite if and only if $\lambda_i(N) + \lambda_{n+1-i}(N) = 2$ for all $i \in \{1, \ldots, n\}$.

Exercise 3.1.4 From [27, Lemma 1.9]. Let $G = (X, E)$ be a connected graph without isolated vertices and N its normalized Laplace matrix. Let $d(G)$ be the diameter of G (see Exercise 1.5.3). We want to find a lower estimate for $\lambda_2(N)$ using $d(G)$.

1. Let $0 \ne f \in \mathbb{K}^X$, such that $D^{\frac{1}{2}} f$ is an eigenvector of N to $\lambda_2(N)$. Let $x_0 \in X$ with $|f(x_0)| = \max_{x \in X} |f(x)|$. Show: There exists a $y_0 \in X$ with $f(x_0) f(y_0) < 0$.
2. Let $x_0, x_1, \ldots, x_n = y_0$ be a path from x_0 to y_0 for the above chosen y_0. Show:

$$\left| f(x_0) - f(y_0) \right|^2 \le k \sum_{j=1}^{k} \left| f(x_j) - f(x_{j-1}) \right|^2.$$

3. Deduce that

$$\langle f, Lf \rangle \ge \frac{\left| f(x_0) - f(y_0) \right|^2}{d(G)}.$$

4. Prove the inequality $\lambda_2(N) \ge \dfrac{1}{d(G) \deg X}$.
5. Let L be the Laplace matrix of G. Use the same method to estimate $\lambda_2(L)$ from below.

Exercise 3.1.5 Let $G = (X, E)$ be a connected graph without isolated vertices and N its normalized Laplace matrix. Define

$$T : \mathbb{K}^X \to \mathbb{K}^X, \quad (Tf)(x) = \frac{1}{\deg x} \sum_{y:\, y \sim x} f(y).$$

Choose a $f \in \mathbb{K}^X$ and define iteratively $f_0 := f$ and $f_k := T f_{k-1}$ for all $k \in \mathbb{N}$.

1. Calculate f_k for $G = K_2$.
2. Let $g_k = D^{\frac{1}{2}} f_k$. Show that $g_k = (I - N) g_{k-1}$ for all $k \in \mathbb{N}$.
3. Let h_i be eigenvectors of N to $\lambda_i(N)$, which form an orthonormal basis of $\mathbb{K}^X$. Define $c_i(k) := \langle h_i, g_k \rangle$. Show:

$$c_i(k) = \big(1 - \lambda_i(N)\big)\, c_i(k - 1) \quad \text{for all } k \text{ and } i.$$

4. Assume that G is not bipartite. Show:
   (a) $c_1(k)$ is constant in k,
   (b) $\lim\limits_{k \to \infty} c_i(k) = 0$ for all $i \geq 2$,
   (c) $\lim\limits_{k \to \infty} f_k = F\mathbb{1}$ with

$$F = \sum_{x \in X} \frac{\deg x}{\deg X} f(x).$$

5. Can the asymptotic behavior of f_k for $k \to \infty$ also be described for bipartite graphs?

Exercise 3.1.6 From the article [62]. Let $G = (X, E)$ be a connected graph without isolated vertices and N be its normalized Laplace matrix. Denote $n := |X|$ and $m := |E|$. We want to show a new estimate for the anticlique number:

$$\alpha(G) \leq 2m \frac{\lambda_n(N) - 1}{\lambda_n(N)\, \deg_{\min} G}. \tag{3.11}$$

For this, let (λ_i, f_i) be eigenpairs of N such that the vectors f_i form an orthonormal basis of $\mathbb{K}^X$ and $\lambda_i \leq \lambda_{i+1}$ for all i.

1. Let $f \in \mathbb{K}^X$. Show:
   (a) $\langle f, Nf \rangle \leq \lambda_n \|f\|^2 - \lambda_n \big|\langle f_1, f \rangle\big|^2$,
   (b) $\langle f, D^{-\frac{1}{2}} A D^{-\frac{1}{2}} f \rangle \geq \lambda_n \big|\langle f_1, f \rangle\big|^2 + (1 - \lambda_n) \|f\|^2$.
2. Show that $f_1 = \pm \frac{1}{\sqrt{2m}} d^{\frac{1}{2}}$ holds.
3. Let $g = D^{-\frac{1}{2}} f$ with $f \in \mathbb{K}^X$. Show:

$$\lambda_n \big|\langle d, g \rangle\big|^2 + 2m(1 - \lambda_n)\langle g, Dg \rangle \leq 4m\, \mathrm{Re} \sum_{e \in E} \overline{g(\iota e)} g(\tau e).$$

4. Let $Y \subset X$ be an anticlique. Show:

$$\lambda_n (\deg Y)^2 + 2m(1 - \lambda_n) \deg Y \leq 0.$$

Hint: Use $g := \mathbb{1}_Y$.

5. Prove the inequality (3.11).

3.2   Nodal Domains on Graphs

Definition 3.2.1 Let $G = (X, E)$ be a connected graph and $f \in \mathbb{R}^X$ a function with $f \not\equiv 0$. A <u>non-empty</u> subset $Y \subset X$ is called a **strong nodal domain** of f, if:

  (i) Y is connected in G (i.e., the induced graph G_Y is connected),

 (ii) for all $x, y \in Y$ it holds that $f(x)f(y) > 0$,

(iii) Y is maximal, i.e., it is not contained in any strictly larger set with properties (i)–(ii).

All $f(y)$ with $y \in Y$ have the same sign, which is called the **sign** of Y.

For $g \in \mathbb{K}^X$ denote

$$\operatorname{supp} g := \{x \in X : g(x) \neq 0\} \quad \text{(the \textbf{support} of } g\text{)}.$$

The following properties are derived directly from the definition:

Proposition 3.2.2 (Properties of Strong Nodal Domains) *Let $G = (X, E)$ be a connected graph and $f \in \mathbb{R}^X$ with $f \not\equiv 0$.*

(a) If Y and Y' are strong nodal domains of f, they are either disjoint or equal.

(b) If Y and Y' are two different strong nodal domains of f with the same sign, there is no edge from Y to Y'.

(c) Every vertex of $\operatorname{supp} f$ *is contained in exactly one strong nodal domain of f.*

Proof

(a) For $Y \cap Y' \neq \varnothing$, the set $Y \cup Y'$ satisfies conditions (i)–(ii) of Definition 3.2.1. From the maximality, it follows that $Y = Y \cup Y' = Y'$.

(b) If there is an edge from Y to Y', then $Y \cup Y'$ is connected and all $f(x)$ with $x \in Y \cup Y'$ have the same sign. According to the maximality, then $Y = Y \cup Y' = Y'$ must hold: contradiction.

(c) Let $x \in X$ with $f(x) \neq 0$ and consider

$$\mathcal{U} := \big\{ U \subset X :\ x \in U \text{ and } U \text{ satisfies (i)--(ii) of Definition 3.2.1} \big\}.$$

Obviously $\{x\} \subset \mathcal{U}$, so $\mathcal{U}$ is not empty. Then

$$Y := \bigcup_{U \in \mathcal{U}} U$$

is a strong nodal domain that contains x. Uniqueness follows from (a).

$\square$

Definition 3.2.3 Let $G = (X, E)$ be a connected graph and $f \in \mathbb{R}^X$ with $f \not\equiv 0$. A non-empty subset $Y \subset X$ is called a **weak nodal domain** of f, if:

  (i) Y is connected in G,
 (ii) for all $x, y \in Y$ it holds that $f(x) f(y) \geq 0$,
(iii) Y is maximal, i.e., it is not contained in any strictly larger set with the two
      above properties.

Definitions 3.2.1 and 3.2.3 differ only in condition (ii), but this affects some important properties.

Proposition 3.2.4 (Properties of Weak Nodal Domains)   *Let $G = (X, E)$ be a connected graph and $f \in \mathbb{R}^X$ with $f \not\equiv 0$.*

(a) If Y is a weak nodal domain of f, then there exists a vertex $y \in Y$ with $f(y) \neq 0$.
    *It follows that for all $y' \in Y$ with $f(y') \neq 0$ the numbers $f(y')$ have the same sign: It is referred to as the **sign** of Y.*
(b) If Y and Y' are two different weak nodal domains of f with $Y \cap Y' \neq \varnothing$, then

  – $f(y) = 0$ for all $y \in Y \cap Y'$,
  – Y and Y' have opposite signs.

(c) Every weak nodal domain includes at least one strong nodal domain.
(d) Every strong nodal domain is contained in exactly one weak nodal domain, and this weak nodal domain has the same sign.

Proof

(a) Proof by contradiction: Assume that $f(y) = 0$ for all $y \in Y$.
    Choose an arbitrary $a \in Y$, then $f(a) = 0$. Since $f \not\equiv 0$, there exists a $b \in X$ with $f(b) \neq 0$. Since G is connected, there is a path $(a = x_0, x_1, \ldots, x_n = b)$ from a to b in G. Choose the smallest k with $f(x_k) \neq 0$. By assumption $x_k \notin Y$.

The set $Y' := Y \cup \{x_1, \ldots, x_k\}$ satisfies conditions (i)–(ii) of Definition 3.2.3 and is strictly larger than Y (since it contains at least one additional element x_k), which contradicts the maximality of Y.

(b) If Y and Y' have the same sign, then $Y \cup Y'$ is connected and satisfies conditions (i)–(ii) of Definition 3.2.3. From the maximality of weak nodal domains, it follows that $Y = Y \cup Y' = Y'$, which contradicts the assumption. Therefore, Y and Y' have opposite signs.

Let $a \in Y$ and $a' \in Y'$ with $f(a) \neq 0$ and $f(a') \neq 0$, then $f(a)f(a') < 0$. Let $x \in Y \cap Y'$. Assume that $f(x) \neq 0$, then $f(x)f(a) > 0$ because $x \in Y$ and $f(x)f(a') > 0$ because $x \in Y'$. Thus, for the product: $f(x)^2 f(a)f(a') > 0$ and therefore $f(a)f(a') > 0$: contradiction.

(c) Let Y be a weak nodal domain. According to (a), there is an $a \in Y$ with $f(y) \neq 0$. Then there is a strong nodal domain $Y' \ni a$ according to Proposition 3.2.2 (c). The set $Y \cup Y'$ is connected and for all $x \in Y \cup Y'$ with $f(x) \neq 0$, $f(x)$ has the same sign as $f(a)$, so $Y \cup Y'$ satisfies conditions (i)–(ii) of Definition 3.2.3 and from the maximality of Y it follows $Y = Y \cup Y'$ and thus $Y' \subset Y$.

(d) Let Y be a strong nodal domain. Consider

$$\mathcal{U} := \{U \subset X : Y \subset U \text{ and } U \text{ satisfies (i)–(ii) of Definition 3.2.3}\};$$

since $Y \in \mathcal{U}$, the set $\mathcal{U}$ is not empty. Then

$$V := \bigcup_{U \in \mathcal{U}} U$$

is a weak nodal domain with $Y \subset U$. Assume that there is another weak nodal domain V' with $Y \subset V'$. According to (b), $f(y) = 0$ for each $y \in Y \subset V \cap V'$, which contradicts the definition of a strong nodal domain.

$\square$

Definition 3.2.5 For a given graph $G = (X, E)$ and for $f \in \mathbb{R}^X$ with $f \neq 0$, we define:

$$\mathcal{N}^{\pm}(f) := \text{the number of strong nodal domains of } f \text{ with the sign } \pm,$$

$$\mathcal{N}_0^{\pm}(f) := \text{the number of weak nodal domains of } f \text{ with the sign } \pm,$$

$$\mathcal{N}(f) := \mathcal{N}^+(f) + \mathcal{N}^-(f) \text{ (the total number of strong nodal domains of } f),$$

$$\mathcal{N}_0(f) := \mathcal{N}_0^+(f) + \mathcal{N}_0^-(f) \text{ (the total number of weak nodal domains of } f).$$

With Proposition 3.2.4(c,d) we obtain the inequalities

$$\mathcal{N}_0^{\pm}(f) \leq \mathcal{N}^{\pm}(f), \qquad \mathcal{N}_0(f) \leq \mathcal{N}(f).$$

Note that $\mathcal{N}(f)$ cannot be arbitrarily large. Obviously $\mathcal{N}(f) \leq |X|$, but this is not the only restriction. For example, a function on a complete graph can have at most 2 strong nodal domains. In fact, the upper bound $|X|$ can only be reached on bipartite graphs, as the following results shows.

Proposition 3.2.6 *Let $G = (X, E)$ be a connected graph, then the following conditions are equivalent:*

(a) There is a $0 \not\equiv f \in \mathbb{R}^X$ with $\mathcal{N}_0(f) = |X|$,
(b) There is a $0 \not\equiv f \in \mathbb{R}^X$ with $\mathcal{N}(f) = |X|$,
(c) G is bipartite.

Proof (a) $\Rightarrow$(b) Let $0 \not\equiv f \in \mathbb{R}^X$ with $\mathcal{N}_0(f) = |X|$. Since there is a vertex x with $f(x) \neq 0$ in every weak nodal domain, each nodal domain consists of only one vertex, in particular f has no zeros. Then every weak nodal domain is a strong nodal domain and $\mathcal{N}(f) = \mathcal{N}_0(f) = |X|$.

(b) $\Rightarrow$(c) Let $0 \not\equiv f \in \mathbb{R}^X$ with $\mathcal{N}(f) = |X|$. Then each strong nodal domain consists of one vertex and f has no zeros. Consider

$$Y := \big\{x \in X : f(x) > 0\big\}, \quad Y' := \big\{x \in X : f(x) < 0\big\},$$

then there are no edges within Y and no edges within Y' (if $a, b \in Y$ with $a \sim b$, then a and b are in the same strong nodal domain, which would then contain at least 2 vertices, analogously for Y'), so all edges go from Y to Y', hence, G is bipartite.

(c) Let G be bipartite with partition components Y and Y'. Define $f := \mathbb{1}_Y - \mathbb{1}_{Y'}$, then $f(x)f(y) < 0$ for all $x \sim y$, i.e., each vertex forms its own strong/weak nodal domain, so $\mathcal{N}_0(f) = |X|$.

$\square$

The following estimate was proved in [89]:

Theorem 3.2.7 (Oren) *Let $G = (X, E)$ be a connected graph with chromatic number $\chi(G)$. For every function $f : X \to \mathbb{R}$ with $f \not\equiv 0$ one has*

$$\mathcal{N}(f) \leq |X| + 2 - \chi(G).$$

For $f \geq 0$ or $f \leq 0$, even $\mathcal{N}(f) \leq |X| + 1 - \chi(G)$ holds.

Proof Denote

$$S := \{x \in X : f(x) = 0\}, \quad N_\pm := \mathcal{N}^\pm(f).$$

First consider the case with $N^+ > 0$ and $N^- > 0$ (i.e., neither $f \geq 0$ nor $f \leq 0$ holds). Let $Y_1^\pm, \ldots, Y_{N_\pm}^\pm$ be the strong nodal domains with the sign $\pm$, which we number so that

$$|Y_1^+| \geq |Y_2^+| \geq \cdots \geq |Y_{N_+}^+|, \qquad |Y_1^-| \geq |Y_2^-| \geq \cdots \geq |Y_{N_-}^-|, \tag{3.12}$$

then we have the disjoint decomposition $X = Y_1^+ \cup \cdots \cup Y_{N_+}^+ \cup Y_1^- \cup \cdots \cup Y_{N_-}^- \cup S$.

Now we iteratively construct a vertex coloring for G. We can first use $|Y_1^+|$ colors for the vertices in Y_1^+ by giving each vertex its own color. Since there are no edges between Y_j^+ with different j, we can use the same colors for all Y_j^+ with $j \geq 2$: Due to (3.12) there are enough colors so that all vertices within Y_j^+ (with a fixed j) have different colors. Now we use another $|Y_1^-|$ colors for the vertices in Y_1^-, then we can similarly use the same colors for each Y_j^- with $j \geq 2$. In addition, we use $|S|$ more colors for the vertices in S. Thus, we have constructed a vertex coloring of G with $|Y_1^+| + |Y_1^-| + |S|$ colors, i.e.

$$\chi(G) \leq |Y_1^+| + |Y_1^-| + |S|.$$

We obviously have

$$\sum_{j=1}^{N_+} |Y_j^+| + \sum_{j=1}^{N_-} |Y_j^-| + |S| = |X|.$$

Since there is at least one vertex in each $Y_j^{\pm}$, it follows

$$|Y_1^+| + (N_+ - 1) + |Y_1^-| + (N_- - 1) + |S| \leq |X|,$$

thus $\mathcal{N}(f) \equiv N_+ + N_- \leq |X| + 2 - \big(|Y_1^+| + |Y_1^-| + |S|\big) \leq |X| + 2 - \chi(G)$.

Now assume that one of the numbers $N_{\pm}$ is zero. W.l.o.g. let $N_- = 0$ (otherwise consider $-f$ instead of f and use $\mathcal{N}(f) = \mathcal{N}(-f)$), i.e. $f \geq 0$. Let $Y_1^+, \ldots, Y_{N_+}^+$ be the strong nodal domains (which then all have the sign $+$), numbered with

$$|Y_1^+| \geq |Y_2^+| \geq \cdots \geq |Y_{N_+}^+|,$$

then we have the disjoint decomposition $X = Y_1^+ \cup \cdots \cup Y_{N_+}^+ \cup S$. Then, as above, we can cover all vertices in all Y_j^+ with $|Y_1^+|$ colors, in addition we use $|S|$ more colors for the vertices in S, thus $\chi(G) \leq |Y_1^+| + |S|$. It follows

$$|Y_1^+| + (N_+ - 1) + |S| \leq \sum_{j=1}^{N_+} |Y_j^+| + |S| = |X|,$$

$$\mathcal{N}(f) \equiv N_+ \leq |X| + 1 - \big(|Y_1^+| + |S|\big) \leq |X| + 1 - \chi(G).$$

$\square$

3.3   Nodal Domains of Eigenfunctions

Now we deal with the nodal domains of eigenfunctions of certain matrices that are induced by G.

Definition 3.3.1 Let $G = (X, E)$ be a graph. A symmetric real $X \times X$ matrix $M = (M_{x,y})_{x,y \in X}$ is a **weighted Laplace matrix for G**, if:

$$M_{x,y} < 0 \text{ for all } x \neq y \text{ with } x \sim y,$$
$$M_{x,y} = 0 \text{ for all } x \neq y \text{ with } x \nsim y.$$

Important: There are no conditions for the diagonal coefficients $M_{x,x}$.

Obviously, the usual Laplace matrix L for G is also a weighted Laplace matrix. The normalized Laplace matrix N provides another example.

Proposition 3.3.2 (Smallest Eigenvalue of Weighted Laplace Matrices) *If M is a weighted Laplace matrix for a connected graph $G = (X, E)$, then $\lambda_{\min}(M)$ is simple, with a positive eigenfunction.*

Proof Let $m := 1 + \max_{x \in X} M_{x,x}$, then the matrix $mI - M$ satisfies the conditions of the strong version of the Perron-Frobenius theorem (Theorem 2.3.3), so the eigenvalue $\lambda_{\max}(mI - M)$ is simple and has a positive eigenfunction. The statement now follows from the equality $\lambda_{\min}(M) = m - \lambda_{\max}(mI - M)$ and

$$\ker\left(M - \lambda_{\min}(M)I\right) = \ker\left(\lambda_{\min}(M)I - M\right)$$
$$= \ker\left((mI - M) - \lambda_{\max}(mI - M)I\right).$$

$\square$

Corollary 3.3.3 *If M is a weighted Laplace matrix for a connected graph $G = (X, E)$ and f is a real eigenfunction of M for $\lambda_{\min}(M)$, then $\mathcal{N}_0(f) = \mathcal{N}(f) = 1$.*

Proof According to Proposition 3.3.2, either $f > 0$ or $f < 0$, so X is the only strong and weak nodal domain. $\square$

We now investigate the number of nodal domains for arbitrary eigenfunctions of weighted Laplace matrices. The following calculation from [48] is used several times and is very useful for the investigation of the nodal domains:

Lemma 3.3.4 (Duval-Reiner Formula) *Let $G = (X, E)$ be a graph with an orientation. Let $M = (M_{x,y})$ be a weighted Laplace matrix for G and denote*

$$M_e := M_{\iota e, \tau e} \equiv M_{\tau e, \iota e} \text{ for } e \in E.$$

Then for all $p \in \mathbb{K}^X$ and $f \in \mathbb{R}^X$ the following identity holds:

$$\langle pf, M(pf) \rangle = \langle |p|^2 f, Mf \rangle - \sum_{e \in E} M_e \, f(\iota e) f(\tau e) \, |p(\iota e) - p(\tau e)|^2.$$

Here $|p|^2$, pf, $|p|^2 f$ are defined by pointwise operations, i.e.

$$|p|^2 : \; X \ni x \mapsto |p(x)|^2 \in \mathbb{R}, \qquad pf : \; X \ni x \mapsto p(x) f(x) \in \mathbb{K},$$

$$|p|^2 f : \; X \ni x \mapsto |p(x)|^2 f(x) \in \mathbb{R}.$$

Proof For each $g \in \mathbb{K}^X$ and each $x \in X$ it holds

$$(Mg)(x) = \sum_{y \in X} M_{x,y} g(y) = - \sum_{y: \, y \sim x} M_{x,y} \big(g(x) - g(y) \big) + W(x) g(x),$$

where

$$W(x) := M_{x,x} + \sum_{y: \, y \sim x} M_{x,y}.$$

With this we obtain

$$\langle pf, M(pf) \rangle = \sum_{x \in X} \overline{(pf)(x)} \big(M(pf) \big)(x)$$

$$= \sum_{x \in X} \overline{p(x)} f(x) \left[- \sum_{y: \, y \sim x} M_{x,y} \big(p(x) f(x) - p(y) f(y) \big) + W(x) p(x) f(x) \right].$$

With the help of

$$p(x) f(x) - p(y) f(y) = p(x) \big(f(x) - f(y) \big) + f(y) \big(p(x) - p(y) \big)$$

we arrive at

$$\langle pf, M(pf) \rangle = \sum_{x \in X} \overline{p(x)} f(x) \left(- p(x) \sum_{y: \, y \sim x} M_{x,y} \big(f(x) - f(y) \big) \right.$$

$$\left. - \sum_{y: \, y \sim x} M_{x,y} f(y) \big(p(x) - p(y) \big) + W(x) p(x) f(x) \right)$$

$$= \sum_{x \in X} |p(x)|^2 f(x) \underbrace{\left(- \sum_{y:\, y \sim x} M_{x,y} \Big(f(x) - f(y) \Big) + W(x) f(x) \right)}_{=(Mf)(x)}$$

$$- \sum_{x \in X} \sum_{y:\, y \sim x} \overline{p(x)}\, f(x) f(y)\, M_{x,y} \Big(p(x) - p(y) \Big)$$

$$= \big\langle |p|^2 f,\, Mf \big\rangle - R$$

$$\text{with } R := \sum_{x \in X} \sum_{y:\, y \sim x} \overline{p(x)}\, f(x) f(y)\, M_{x,y} \Big(p(x) - p(y) \Big).$$

In the expression for R each edge $e \in E$ appears twice: as $(x,y) = (\iota e, \tau e)$ and as $(x,y) = (\tau e, \iota e)$, while in both cases $M_{x,y} = M_e$. Thus

$$R = \sum_{e \in E} \left(\overline{p(\iota e)}\, f(\iota e) f(\tau e)\, M_e \Big(p(\iota e) - p(\tau e) \Big) \right.$$

$$\left. + \overline{p(\tau e)}\, f(\tau e) f(\iota e)\, M_e \Big(p(\tau e) - p(\iota e) \Big) \right)$$

$$\equiv \sum_{e \in E} f(\iota e) f(\tau e) M_e \Big(\overline{p(\tau e)} - \overline{p(\iota e)} \Big) \Big(p(\tau e) - p(\iota e) \Big)$$

$$= \sum_{e \in E} M_e\, f(\iota e) f(\tau e)\, \big| p(\iota e) - p(\tau e) \big|^2.$$

$$\square$$

Now we prove a central result about the nodal domains. This version comes from the article [43].

Theorem 3.3.5 (Nodal Theorem for Eigenfunctions) *Let $G = (X, E)$ be a connected graph and M a weighted Laplace matrix for G. Let*

$$\lambda_1(M) < \lambda_2(M) \leq \cdots \leq \lambda_{|X|}(M)$$

be the eigenvalues of M. Let $k \in \{1, \ldots, |X|\}$ and denote:

$$\underline{k} := \min \big\{ j : \lambda_j(M) = \lambda_k(M) \big\}, \qquad \overline{k} := \max \big\{ j : \lambda_j(M) = \lambda_k(M) \big\},$$

then every real eigenfunction of M to $\lambda_k(M)$ has

(a) at most $\overline{k}$ strong nodal domains,
(b) at most $\underline{k}$ weak nodal domains.

Proof Denote $n := |X|$ and $\lambda_j := \lambda_j(M)$ for $j \in \{1, \ldots, n\}$. Let f be a real eigenfunction of M to λ_k and w.l.o.g. let $\|f\| = 1$. Then we can construct an orthonormal basis $(f_1, \ldots, f_n)$ of $\mathbb{K}^X$ such that

$$Mf_j = \lambda_j f_j \text{ for all } j \in \{1, \ldots, n\} \quad \text{and} f_{\underline{k}} = f.$$

The last condition will be important at the end of the proof.

Let $X_1, \ldots, X_\ell$ be either the strong nodal domains of f or the weak nodal domains of f (a case distinction will be made later). For $i \in \{1, \ldots, \ell\}$ define $g_i \in \mathbb{R}^X$ by

$$g_i : X \ni x \mapsto \begin{cases} f(x), & x \in X_i, \\ 0, & \text{otherwise.} \end{cases}$$

For each i one has $g_i \not\equiv 0$. In addition, the functions g_i have pairwise disjoint supports: In the case of strong nodal domains, X_i are pairwise disjoint, for weak nodal domains this follows from Proposition 3.2.4 (b). Thus $\mathrm{span}\{g_1, \ldots, g_\ell\}$ is an ℓ-dimensional subspace, so there exists some $g \in \mathrm{span}\{g_1, \ldots, g_\ell\}$ with $\|g\| = 1$ and $g \perp f_j$ for all $j \in \{1, \ldots, \ell - 1\}$. By construction

$$g = \sum_{i=1}^{\ell} p_i g_i \quad \text{with } p_i \in \mathbb{K}. \tag{3.13}$$

Write g as a linear combination of the eigenfunctions f_j:

$$g = \sum_{j=1}^{n} \langle f_j, g \rangle f_j \equiv \sum_{j=\ell}^{n} \langle f_j, g \rangle f_j,$$

$$\|g\|^2 = \sum_{j=\ell}^{n} |\langle f_j, g \rangle|^2, \quad Mg = \sum_{j=\ell}^{n} \lambda_j \langle f_j, g \rangle f_j,$$

then it follows

$$\langle g, Mg \rangle = \sum_{j=\ell}^{n} \lambda_j |\langle f_j, g \rangle|^2 \geq \sum_{j=\ell}^{n} \lambda_\ell |\langle f_j, g \rangle|^2 = \lambda_\ell \|g\|^2 = \lambda_\ell. \tag{3.14}$$

Define $p \in \mathbb{K}^X$ by

$$p : X \ni x \mapsto \begin{cases} p_i, & x \in X_i \cap \mathrm{supp}\, f, \\ 0, & \text{otherwise,} \end{cases}$$

with the constants p_i from (3.13), then $g = pf$. Now we use the Duval-Reiner formula (Lemma 3.3.4):

$$\langle g, Mg \rangle = \langle |p|^2 f, Mf \rangle - \sum_{e \in E} r_e \quad \text{with } r_e := M_e \, f(\iota e) f(\tau e) \, |p(\iota e) - p(\tau e)|^2.$$

We have $Mf = \lambda_k f$, so

$$\langle |p|^2 f, Mf \rangle = \lambda_k \langle |p|^2 f, f \rangle = \lambda_k \|pf\|^2 \equiv \lambda_k \|g\|^2 = \lambda_k,$$

and from this it follows

$$\langle g, Mg \rangle = \lambda_k - \sum_{e \in E} r_e. \tag{3.15}$$

Let $e \in E$, then $M_e < 0$ and we are in one of the following three cases:

(i) $f(\iota e) = 0$ or $f(\tau e) = 0$. Then $r_e = 0$.
(ii) $f(\iota e) f(\tau e) > 0$. Then ιe and τe are in the same nodal domain:

$$\iota e, \tau e \in X_i \cap \operatorname{supp} f \text{ for some } i \in \{1, \ldots, \ell\},$$

so $p(\iota e) = p(\tau e) = p_i$ and it follows $r_e = 0$.
(iii) $f(\iota e) f(\tau e) < 0$. In this case $r_e \geq 0$.

Overall, from (3.15) it follows $\langle g, Mg \rangle \leq \lambda_k$. In (3.14) we have shown $\langle g, Mg \rangle \geq \lambda_\ell$, so $\lambda_\ell \leq \lambda_k$ and thus $\ell \leq \overline{k}$. With this we have proven statement (a) about the number of strong nodal domains.

As we now deal with the (much more complex) proof of (b), we assume that X_i are the weak nodal domains of f. We conduct a proof by contradiction: Assume that $\ell > \underline{k}$. This implies $\lambda_\ell \geq \lambda_k$. Since we have also proven $\lambda_\ell \leq \lambda_k$ above, it follows $\lambda_\ell = \lambda_k$, and from the previous calculations it follows $\langle g, Mg \rangle = \lambda_k$.

Let $F := \{f_1, \ldots, f_{\ell-1}\}^\perp$ and denote by M' the restriction of M to F. Then $M' : F \to F$ is self-adjoint with $\lambda_{\min}(M') = \lambda_\ell$. By construction we have $g \in F$ and $\langle g, M'g \rangle \equiv \langle g, Mg \rangle = \lambda_\ell \equiv \lambda_\ell \|g\|^2$. From this it follows by the min-max principle $Mg \equiv M'g = \lambda_\ell g$ (Corollary 2.1.3). With (3.15) one obtains

$$\sum_{e \in E} r_e = 0.$$

In the above case distinction (i)–(iii) we have seen that $r_e \geq 0$ for all $e \in E$, so

$$r_e = 0 \text{ for all } e \in E.$$

For the rest of the proof we will introduce a local definition. Let $i, j \in \{1, \ldots, \ell\}$ with $i \neq j$. We say that X_i is *adjacent* to X_j (and write $X_i \sim X_j$), if there is an edge $e = \{a, b\} \in E$ such with $a \in X_i$ and $b \in X_j$ with $f(b) \neq 0$. (It is easy to see that $X_i \sim X_j$ is equivalent to $X_j \sim X_i$, but this is irrelevant for the following discussion.) We now prove two auxiliary statements (H1) and (H2) about adjacent nodal domains.

(H1) Let $i \in \{1, \ldots, \ell\}$ with $p_i \neq 0$ and let $j \neq i$ with $X_i \sim X_j$. Then $p_i = p_j$.

Proof of (H1) Let $e = \{a, b\} \in E$ be an edge with $a \in X_i$ and $b \in X_j$, such that $f(b) \neq 0$. W.l.o.g, let $a = \iota e$ and $b = \tau e$ (otherwise we choose the opposite orientation of the edge). Now consider two cases.

Case 1: $f(a) \neq 0$. Then $f(\iota e)f(\tau e) \equiv f(a)f(b) \neq 0$ and from $r_e = 0$ it follows that $p(\iota e) = p(\tau e)$ and thus $p_i = p_j$.

Case 2: $f(a) = 0$. We denote

$$Y := \{y \in X : y \sim a \text{ and } y \notin X_i\},$$

then obviously $b \in Y$.

We can w.l.o.g assume that X_i has the positive sign, then $f(y) < 0$ for all $y \in Y$ (from $f(y) \geq 0$ it would follow that $y \in X_i$ due to the maximality of X_i). Then $Y \cup \{a\}$ is connected, with $f(y) \leq 0$ for all $y \in Y \cup \{a\}$, so $Y \cup \{a\}$ is contained in a weak nodal domain. Since $b \in Y \cup \{a\}$ and $b \in X_j$, we have $Y \cup \{a\} \subset X_j$. Then it holds $g(y) = p_j f(y)$ for all $y \in Y$.

Since both f and g are eigenfunctions to λ_k, also

$$h := f - \frac{1}{p_i} g$$

is an eigenfunction to λ_k. We have $h = 0$ on X_i, in particular $h(a) = 0$. According to the above calculation, it holds

$$h(y) = \left(1 - \frac{p_j}{p_i}\right) f(y) \text{ for all } y \in Y.$$

From $h \in \ker(M - \lambda_k I)$ it follows in particular

$$0 = \lambda_k h(a) = (Mh)(a) = \sum_{y \in X} M_{a,y} h(y),$$

where all summands with $y \notin Y$ are zero, so

$$0 = \sum_{y \in Y} M_{a,y} h(y) = \left(1 - \frac{p_j}{p_i}\right) \sum_{y \in Y} M_{a,y} f(y).$$

For all $y \in Y$, $M_{a,y} < 0$ and $f(y) < 0$, so the sum is positive. It follows

$$1 - \frac{p_j}{p_i} = 0,$$

and thus $p_i = p_j$. We have proven (H1).

The second auxiliary statement is:

(H2) Let $\varnothing \neq \mathcal{I} \subsetneq \{1, \ldots, \ell\}$. Then for some $i \in \mathcal{I}$ and $j \in \{1, \ldots, \ell\} \setminus \mathcal{I}$ one has $X_i \sim X_j$.

Proof of (H2) Let $S := \bigcup_{i \in \mathcal{I}} X_i$. According to Proposition 3.2.4 (a) there is a vertex $a \in S$ with $f(a) \neq 0$. By assumption one has $S \neq X$, so there is some $j_0 \in \{1, \ldots, \ell\} \setminus \mathcal{I}$ with $X_{j_0} \not\subset S$. Thanks to Proposition 3.2.4(a) there is a vertex $b \in X_{j_0}$ with $f(b) \neq 0$, then $b \notin S$ by Proposition 3.2.4(d).

Since G is connected, there exists a path $(a = x_0, x_1, \ldots, x_m = b)$. Choose:

- the smallest $q \in \{0, \ldots, m\}$ with $x_q \notin S$ and $f(x_q) \neq 0$; obviously $q \geq 1$,
- the largest $r \in \{0, \ldots, q-1\}$ with $f(x_r) \neq 0$,

then there exists an index $i \in \mathcal{I}$ with $x_r \in X_i$ and an index $j \in \{1, \ldots, \ell\} \setminus \mathcal{I}$ with $x_q \in X_j$.

If $r = q-1$, then $\{x_r, x_q\}$ is an edge from X_i to X_j with $f(x_q) \neq 0$, so $X_i \sim X_j$. If $r \leq q-2$, $f(x_t) = 0$ for all $t \in \{r+1, \ldots, q-1\}$, so these x_t belong to $X_i \cap X_j$. Then $\{x_{q-1}, x_q\}$ is an edge from X_i to X_j with $f(x_q) \neq 0$ and $X_i \sim X_j$. The statement (H2) is proven.

Now we continue with the proof of the theorem. To recall

$$g = \sum_{i=1}^{\ell} p_i g_i, \quad p_i \in \mathbb{K}, \quad \|g\| = 1,$$

so there exists an index $i \in \{1, \ldots, \ell\}$ with $p_i \neq 0$. According to (H2) there is an index $j_1 \in \{1, \ldots, \ell\} \setminus \{i\}$ with $X_i \sim X_{j_1}$, then $p_{j_1} = p_i$ in virtue of (H1). Then there exists an index $j_2 \in \{1, \ldots, \ell\} \setminus \{i, j_1\}$ with $X_i \sim X_{j_2}$ or $X_{j_1} \sim X_{j_2}$, so $p_{j_2} = p_{j_1} = p_i$, etc. In the end, we have $p_j = p_i$ for all j, so $g = p_i f$ with $p_i \in \mathbb{K} \setminus \{0\}$. Thanks to the choice of g, we have $\langle f_j, g \rangle = 0$ for all $j \in \{0, \ldots, \ell-1\}$ and thus also $\langle f_j, f \rangle = 0$ for all $j \in \{0, \ldots, \ell-1\}$. As a reminder: we have $f_k = f$, and $\ell > k$ by assumption. For $j := k$, we then get $f = 0$: a contradiction to the definition of an eigenfunction. Therefore, $\ell \leq k$ must hold.

This concludes the proof of the nodal theorem 3.3.5. $\qquad\qquad\square$

Corollary 3.3.6 (Nodal Domains for the Second Smallest Eigenvalue) *Let $G = (X, E)$ be a connected graph with at least one edge and M a weighted Laplace matrix on G. Let g be a real eigenfunction to $\lambda_2(M)$, then:*

(a) $\mathcal{N}_0(g) = 2$ and the two weak nodal domains have opposite signs,
(b) $2 \leq \mathcal{N}(g) \leq \overline{2}$, with at least one strong nodal domain for each sign.

Proof According to Proposition 3.3.2, there is an eigenfunction $f > 0$ for $\lambda_1(M)$, and $\lambda_1(M) < \lambda_2(M)$, in particular, $\underline{2} = 2$.

From $\langle f, g \rangle = 0$ it follows that g must take negative and positive values, so g has at least one strong positive nodal domain and one negative strong nodal domain. This results in the lower bound in (b), with the upper bound following from the nodal theorem. If $\lambda_2(M)$ is simple, then $\overline{2} = 2$.

Similarly, g has at least one positive weak nodal domain and at least one negative weak nodal domain, so $\mathcal{N}_0(g) \geq 2$, and from the nodal theorem it follows that $\mathcal{N}_0(g) \leq \underline{2} = 2$. This also proves (a). $\qquad\square$

In Exercise 3.3.1 it is shown that the lower bound in Corollary 3.3.6(b) is attained by at least one eigenfunction [106] (this property will play a role in Sect. 4.5):

Proposition 3.3.7 *Let $G = (X, E)$ be a connected graph with at least two vertices, M a weighted Laplace matrix on G and g a real eigenfunction of M to $\lambda_2(M)$ with minimal support, i.e., each eigenfunction g' for $\lambda_2(M)$ with $\operatorname{supp} g' \subset \operatorname{supp} g$ satisfies $\operatorname{supp} g' = \operatorname{supp} g$. Then g has exactly one positive strong nodal domain and exactly one negative strong nodal domain.*

The case of bipartite graphs can be examined in more detail [95].

Theorem 3.3.8 (Roth: Largest Eigenvalue on Bipartite Graphs) *Let $G = (X, E)$ be a connected bipartite graph with partition components Y and Y' and M a weighted Laplace matrix for G, then:*

(a) *There is an eigenfunction g of M for $\lambda_{\max}(M)$ with $g > 0$ on Y and $g < 0$ on Y',*
(b) *$\lambda_{\max}(M)$ is a simple eigenvalue.*

Proof Consider an orientation on G. For each $f \in \mathbb{K}^X$, it follows from the Duval-Reiner formula (Lemma 3.3.4) with $(p, f) := (f, \mathbb{1})$ that

$$\langle f, Mf \rangle = \langle |f|^2 \mathbb{1}, M\mathbb{1} \rangle - \sum_{e \in E} M_e |f(\iota e) - f(\tau e)|^2$$

$$= -\sum_{e \in E} M_e |f(\iota e) - f(\tau e)|^2 + \sum_{x \in X} W(x) |f(x)|^2 \qquad (3.16)$$

$$\text{with } W : X \ni x \mapsto \sum_{y \in X} M_{x,y}.$$

As a reminder, $M_e < 0$ for all $e \in E$. Now let f be a real eigenfunction of M for the eigenvalue $\lambda_{\max}(M)$, then $\langle f, Mf \rangle = \lambda_{\max}(M)\|f\|^2$. Define $g : X \to \mathbb{R}$ by

$$g : X \ni x \mapsto \begin{cases} |f(x)|, & x \in Y, \\ -|f(x)|, & x \in Y'. \end{cases}$$

For each $x \in X$, $|g(x)| = |f(x)|$, in particular $\|g\| = \|f\|$, and $\langle g, Mg \rangle \leq \lambda_{\max}(M)\|g\|^2$ by the min-max principle. Since every edge in G goes from Y to Y', for each $e \in E$ we have

- either $g(\iota e) = |f(\iota e)|$ and $g(\tau e) = -|f(\tau e)|$,
- or $g(\iota e) = -|f(\iota e)|$ and $g(\tau e) = |f(\tau e)|$,

so $|g(\iota e) - g(\tau e)| = |f(\iota e)| + |f(\tau e)| \geq |f(\iota e) - f(\tau e)|$. By substituting these inequalities into (3.16) one obtains

$$\langle g, Mg \rangle \geq \langle f, Mf \rangle \equiv \lambda_{\max}(M)\|f\|^2 \equiv \lambda_{\max}(M)\|g\|^2.$$

In total, we have $\langle g, Mg \rangle = \lambda_{\max}(M)\|g\|^2$, so g is an eigenfunction of M for $\lambda_{\max}(M)$, and by construction we have $g \geq 0$ on Y and $g \leq 0$ on Y'. It remains to show that g has no zeros.

Assume that $g(x) = 0$ for some $x \in X$. Consider two cases:

- If $x \in Y$, then all $y \in X$ with $y \sim x$ are in Y', so $g(y) \leq 0$. Then from

$$0 = \lambda_{\max}(M)g(x) = M_{x,x}\underbrace{g(x)}_{=0} + \sum_{y \sim x}\underbrace{M_{x,y}}_{<0}\underbrace{g(y)}_{\leq 0},$$

  it follows that $g(y) = 0$ for all $y \sim x$.
- If $x \in Y'$, one can show in a completely analogous way that $g(y) = 0$ for all $y \sim x$.

So in both cases $g(y) = 0$ for all $y \sim x$. Since G is connected, one obtains iteratively $g \equiv 0$, which is obviously wrong. Therefore, g has no zeros. This proves (a).

To prove (b) we use the nodal theorem 3.3.5. Let $n := |X|$ and assume that $\lambda_n(M) = \lambda_{\max}(M)$ is not simple, then $\lambda_{n-1}(M) = \lambda_n(X)$. Let g be the eigenfunction of M with respect to $\lambda_n(M)$ constructed in (a), then according to the nodal theorem it has at most $\underline{n} \leq n - 1$ weak nodal domains. But g obviously has exactly n weak nodal domains: contradiction. $\qquad\square$

By combining with Proposition 3.3.2 we obtain:

Corollary 3.3.9 *If M is a weighted Laplace matrix for a connected bipartite graph, then the smallest and largest eigenvalues of M are simple.*

Remark 3.3.10 Similar to the Cheeger inequality, the nodal theorem for the eigenfunctions of graph matrices (as well as the investigation of sign-changing edges in the next sections) was motivated by known results about partial differential equations.

Let $\Omega \subset \mathbb{R}^n$ be a non-empty connected bounded open set with smooth boundary $\partial\Omega$. For each $x \in \overline{\Omega}$ let $A(x)$ be a $n \times n$ real symmetric matrix, such that $x \mapsto A(x)$ is continuously differentiable and $\lambda_1\big(A(x)\big) > 0$ for all $x \in \overline{\Omega}$. Furthermore, let $\rho : \overline{\Omega} \mapsto \mathbb{R}$ be a continuous function. Consider the differential expression

$$P : f \mapsto -\sum_{j,k=1}^{n} \frac{\partial}{\partial x_j}\left(A_{j,k}\,\frac{\partial f}{\partial x_k}\right) + \rho f.$$

The numbers $\lambda \in \mathbb{R}$, for which the boundary value problem

$$Pf = \lambda f \text{ in } \Omega, \quad f|_{\partial\Omega} = 0,$$

has a twice continuously differentiable solution $f \not\equiv 0$, are referred to as the *eigenvalues* of the boundary value problem and the corresponding solutions f are called *eigenfunctions*. The eigenvalues form a non-decreasing sequence $(\lambda_k)_{k\in\mathbb{N}}$ (each eigenvalue appears as often as the dimension of the corresponding solution space indicates) with $\lim_{k\to\infty} \lambda_k = \infty$. If f_k is an eigenfunction for λ_k, then the set of its zeros is referred to as the *nodal set* of f_k. The connected components of $\Omega \setminus$ (nodal set of f_k) are called *nodal domains* of f_k and their number we denote by $\mathcal{N}(f_k)$. The *Courant nodal domain theorem* from 1923 states that for each $k \in \mathbb{N}$ and any choice of the eigenfunction f_k for λ_k one has $\mathcal{N}(f_k) \leq k$, see [34, Chapter 6, §6].

The one-dimensional case ($n = 1$) was studied much earlier: Already in 1833, Sturm observed that $\mathcal{N}(f_k) = k$ holds for all k (in this case Ω is an interval, so f_k has exactly $k - 1$ zeros within Ω). The corresponding theorems can now be found in most books on ordinary differential equations, see e.g. [112, §27]. In 1956, Pleijel [91] discovered that the dimension really plays a role: For $n \geq 2$, the equality $\mathcal{N}(f_k) = k$ can only hold for finitely many k and there is even a constant $c < 1$, such that $\mathcal{N}(f_k) \leq ck$ holds for all sufficiently large k. These and other similar results also hold in much more general situations (other differential expressions and boundary conditions, operators on manifolds, etc.). It should be noted that differential operators and graphs can be considered in a uniform abstract way, see e.g. [68]. The study of the nodal domains of eigenfunctions remains a very active field of research [25, 121].

Many further parallels between graph matrices and differential operators can be found e.g. in the review article by Sunada [104] and in the books by Barlow [12], Grigor'yan [60] and Keller et al. [69].

Further results on the eigenfunctions of various graph matrices can be found e.g. in the books by Cvetković et al. [42] and Biyikoğlu et al. [19].

Exercise 3.3.1 Let $G = (X, E)$ be a connected graph and M a weighted Laplace matrix for G. Further, let $f \in \mathbb{R}^X$ be an eigenvector for the eigenvalue $\lambda_2(M)$ with minimal support, i.e., there is no eigenvector for $\lambda_2(M)$ whose support is strictly contained in the support of f. Define

$$X_+ := \{x \in X : f(x) > 0\}, \quad X_- := \{x \in X : f(x) < 0\}.$$

The two $X_\pm$ are non-empty, and we want to show:

$$\text{The induced graphs } G_{X_+} \text{ and } G_{X_-} \text{ are connected.} \tag{3.17}$$

This is obviously equivalent to $\mathcal{N}(f) = 2$ (and proves Proposition 3.3.7).

For non-empty subsets $S, T \subset X$ and arbitrary $g \in \mathbb{R}^X$, we denote by $M_{S,T}$ the $S \times T$ matrix $(M_{x,y})_{(x,y) \in S \times T}$ and by g_S the restriction of g to S.

We conduct a proof by contradiction: Assume (3.17) is false. W.l.o.g. we assume that $\lambda_2(M) = 0$ and that G_{X_+} is not connected. Then there is a connected component $S \subset X_+$, which is not connected to the rest $T := X_+ \setminus S$. We denote $R := X_-$. Show:

1. $M_{S,T} = 0$ and $M_{T,S} = 0$,
2. $M_{S,S} f_S + M_{S,R} f_R = 0$ and $M_{T,T} f_T + M_{T,R} f_R = 0$.
3. It holds $M_{S,S} f_S \leq 0$ and $M_{T,T} f_T \leq 0$ (component-wise).

Let $h > 0$ be an eigenvector of M for $\lambda_1(M)$. Define $g \in \mathbb{R}^X$ by

$$g(x) = \begin{cases} f(x), & x \in S, \\ -\alpha f(x), & x \in T, \\ 0, & \text{otherwise,} \end{cases} \qquad \text{with } \alpha := \frac{\langle h_S, f_S \rangle}{\langle h_T, f_T \rangle}.$$

Show:

4. $\langle h_T, f_T \rangle \neq 0$, thus α is well-defined.
5. $\alpha > 0$,
6. $\langle h, g \rangle = 0$,
7. $\langle g, Mg \rangle \leq 0$,
8. $Mg = 0$.
9. Find the contradiction.

Exercise 3.3.2 From the article [20], see also [19, Chapter 3.6].

Let $G = (X, E)$ be a graph. Let $Y \subset X$, such that the induced graph G_Y is bipartite and $|Y|$ is maximal. Show: For every function $f \in \mathbb{R}^X$ with $f \not\equiv 0$ it holds that $\mathcal{N}(f) \leq |Y|$.

Hint: Choose a vertex in each strong nodal domain of f.

Exercise 3.3.3 From the article [54], see also [19, Chapter 3.4].

Let M be a weighted Laplace matrix for a connected graph $G = (X, E)$ with $n := |X|$. We want to show that there is an orthonormal basis $(f_1, \ldots, f_n)$ of $\mathbb{R}^X$ such that for each $k \in \{1, \ldots, n\}$ it holds $M f_k = \lambda_k(M) f_k$ and $\mathcal{N}(f_k) \leq k$.

The eigenfunctions f_k are constructed iteratively.

1. Show that any normalized eigenfunction of M for $\lambda_1(M)$ can be chosen as f_1.
2. Let $k \geq 2$, such that the functions $f_1, \ldots, f_{k-1}$ have already been constructed. Show: There exists a normalized eigenfunction f for $\lambda_k(M)$ with $f \perp f_j$ for all $j \in \{1, \ldots, k-1\}$.
   (a) Let $\mathcal{N}(f) \leq k$. Show that one can take $f_k := f$.
   (b) Let $\mathcal{N}(f) \geq k+1$. Let $X_1, X_2, \ldots$ be the strong nodal domains of f and consider the functions $g_i := \mathbb{1}_{X_i} f$. Show:
       i. There exists a normalized function $g \in \mathrm{span}\{g_1, \ldots, g_k\}$ such that $g \perp f_j$ for all $j \in \{1, \ldots, k-1\}$ and $\langle g, Mg \rangle = \lambda_k(M)$.
          Hint: Use the first steps in the proof of the nodal theorem 3.3.5.
       ii. The function g is an eigenfunction of M for $\lambda_k(M)$.
       iii. It holds $\mathcal{N}(g) \leq k$.
       iv. One can take $f_k := g$.

3.4 Sign Changes and Betti Number

Definition 3.4.1 Let $G = (X, E)$ be a connected graph and $f \in \mathbb{R}^X$ a function <u>without zeros</u>, i.e., $f(x) \neq 0$ for all $x \in X$. A **sign-changing edge** of f is an edge $e \in E$ with $f(\iota e) f(\tau e) < 0$ (where the orientation is freely chosen). The number of sign-changing edges of f is denoted by $\nu(f)$.

The number $\nu(f)$ is related to the number of nodal domains $\mathcal{N}(f)$ of f as follows:

Proposition 3.4.2 (Sign-Changing Edges and Nodal Domains) *Let $G = (X, E)$ be a connected graph and $f \in \mathbb{R}^X$ a function without zeros, then*

$$\mathcal{N}(f) \geq |X| - |E| + \nu(f).$$

Proof Let $G' = (X, E')$ be the graph obtained by removing the $\nu(f)$ sign-changing edges of f, then $|E'| = |E| - \nu(f)$. Let $G_i = (X_i, E_i)$ with $i \in \{1, \ldots, p\}$ be the connected components of G'. We claim that each X_i is a strong nodal domain of f. By construction, X_i is connected in G and f has a constant sign on X_i. Assume that X_i is not maximal and let Y be a strong nodal domain with $X_i \subsetneq Y$. Let $x \in X_i$ and $y \in Y \setminus X_i$, then $y \in X_j$ for some $j \neq i$. W.l.o.g., let Y be a positive nodal domain, then $f(x) > 0$ and $f(y) > 0$. Then there is a path $(x = a_0, a_1, \ldots, a_n = y)$ in G with $f(a_j) > 0$ for all j. This path cannot be contained in G' (since it runs between two different connected components of G'), so it must necessarily go over one of the sign-changing edges, but then $f(a_j) < 0$ for some $j \in \{1, \ldots, n-1\}$: contradiction. Thus, X_i are the strong nodal domains of f and $p = \mathcal{N}(f)$.

Since all G_i are connected, $|E_i| \geq |X_i| - 1$ (Corollary 1.1.22) and

$$|E| - \nu(f) = |E'| = \sum_{i=1}^{p} |E_i| \geq \sum_{i=1}^{p} \left(|X_i| - 1 \right) = \sum_{i=1}^{p} |X_i| - p = |X| - \mathcal{N}(f).$$

$$\square$$

The following estimate for $\nu(f)$ was found in [14].[1]

Theorem 3.4.3 (Berkolaiko: Sign-Changing Edges of Eigenfunctions) *Let $G = (X, E)$ be a connected graph and M a weighted Laplace matrix for G. Let $k \in \{1, \ldots, |X|\}$ and f a zero-free real eigenfunction of M for $\lambda_k(M)$. Then*

$$\underline{k} - 1 \leq \nu(f) \leq \underline{k} - 1 + \beta(G),$$

where $\underline{k}$ is as in the nodal theorem 3.3.5 and

$$\beta(G) := |E| - |X| + 1$$

is the so-called **first Betti number** *of G. In particular, $\nu(f) = \underline{k} - 1$, when G is a tree.*

Proof Let $\lambda_k := \lambda_k(M)$. For a self-adjoint linear mapping B, we denote by $N_-(B)$ the number of negative eigenvalues of B. Let $g \in \mathbb{K}^X$, then the Duval-Reiner formula (Lemma 3.3.4) can be used,

$$\langle gf, M(gf) \rangle = \langle |g|^2 f, Mf \rangle - \sum_{e \in E} M_e \, f(\iota e) f(\tau e) \left| g(\iota e) - g(\tau e) \right|^2.$$

We have $\langle |g|^2 f, Mf \rangle = \langle |g|^2 f, \lambda_k f \rangle = \lambda_k \langle gf, gf \rangle$, so

$$\langle gf, (M - \lambda_k I)(gf) \rangle = \sum_{e \in E} (-M_e) \, f(\iota e) f(\tau e) \left| g(\iota e) - g(\tau e) \right|^2. \tag{3.18}$$

Consider the following linear mappings:

$$\Phi : \mathbb{K}^X \to \mathbb{K}^X, \quad (\Phi g)(x) := f(x)g(x),$$

$$T : \mathbb{K}^E \to \mathbb{K}^E, \quad (Tu)(e) := t(e)u(e) \text{ with } t(e) := (-M_e) \, f(\iota e) f(\tau e).$$

---

[1] Note for the instructors: If you are planning to cover Sect. 3.5, you can treat Lemma 3.5.3 already as part of Theorem 3.4.3.

The eigenvalues of T are $t(e)$ with $e \in E$ (since T is given by a diagonal matrix), and an edge $e \in E$ is sign-changing for f if and only if $t(e) < 0$. Thus, $\nu(f) = N_-(T)$.

The identity (3.18) has the form $\langle \Phi g, (M - \lambda_k I)\Phi g \rangle = \langle \vec{\nabla} g, T \vec{\nabla} g \rangle$, i.e.

$$\langle g, \Phi^*(M - \lambda_k I)\Phi g \rangle = \langle g, \vec{\nabla}^* T \vec{\nabla} g \rangle \text{ for all } g \in \mathbb{K}^X. \tag{3.19}$$

By the min-max principle it follows that $\Phi^*(M - \lambda_k I)\Phi$ and $\vec{\nabla}^* T \vec{\nabla}$ have the same eigenvalues, in particular $N_-\big(\Phi^*(M - \lambda_k I)\Phi\big) = N_-(\vec{\nabla}^* T \vec{\nabla})$. The function f has no zeros, so Φ is bijective. By Sylvester's law of inertia (Corollary 2.1.11) one has

$$N_-\big(\Phi^*(M - \lambda_k I)\Phi\big) = N_-(M - \lambda_k I) = \underline{k} - 1$$

and thus $\underline{k} - 1 = N_-(\vec{\nabla}^* T \vec{\nabla})$.

Since $\vec{\nabla} : \mathbb{K}^X \to \operatorname{ran} \vec{\nabla}$ is surjective, it will be convenient to use the natural embedding

$$\Theta : \operatorname{ran} \vec{\nabla} \ni v \mapsto v \in \mathbb{K}^E$$

For all $u \in \mathbb{K}^X$ it holds

$$\langle u, \vec{\nabla}^* T \vec{\nabla} u \rangle = \langle \vec{\nabla} u, T \vec{\nabla} u \rangle = \langle \Theta \vec{\nabla} u, T \Theta \vec{\nabla} u \rangle = \langle u, \vec{\nabla}^* \Theta^* T \Theta \vec{\nabla} u \rangle,$$

and by Corollary 2.1.10 we have

$$\begin{aligned}
\underline{k} - 1 &= N_-\big(\vec{\nabla}^* T \vec{\nabla}\big) \\
&= \max \left\{ \dim U : U \subset \mathbb{K}^X \text{ with } \langle u, \vec{\nabla}^* T \vec{\nabla} u \rangle < 0 \text{ for all } u \in U \setminus \{0\} \right\} \\
&= \max \left\{ \dim U : U \subset \mathbb{K}^X \text{ with } \langle u, \vec{\nabla}^* \Theta^* T \Theta \vec{\nabla} u \rangle < 0 \text{ for all } u \in U \setminus \{0\} \right\} \\
&= N_-(\vec{\nabla}^* \Theta^* T \Theta \vec{\nabla}) = N_-(\Theta^* T \Theta),
\end{aligned}$$

where the last equality is satisfied by Sylvester's inertia theorem (Corollary 2.1.11). We then have

$$\dim \ker \vec{\nabla} = \dim \left\{ f \in \mathbb{K}^X : f(\iota e) = f(\tau e) \text{ for all } e \in E \right\} = \dim \mathbb{K}\mathbf{1} = 1,$$

$$\dim \operatorname{ran} \vec{\nabla} = \dim \mathbb{K}^X - 1 = |X| - 1,$$

$$\beta(G) = \dim \mathbb{K}^E - \dim \operatorname{ran} \vec{\nabla}. \tag{3.20}$$

By the interlacing principle (Corollary 2.1.8), the eigenvalues of the mappings

$$T : \mathbb{K}^E \to \mathbb{K}^E, \qquad \Theta^* T \Theta : \operatorname{ran} \vec{\nabla} \to \operatorname{ran} \vec{\nabla}$$

satisfy the inequalities

$$\lambda_k(T) \le \lambda_k(\Theta^* T \Theta) \le \lambda_{k+\beta(G)}(T) \text{ for all } k \in \{1, \ldots, |X|-1\}. \qquad (3.21)$$

For all $k \le N_-(\Theta^* T \Theta)$, $\lambda_k(\Theta^* T \Theta) < 0$, and from the left inequality in (3.21) it follows $\lambda_k(T) < 0$. Thus $N_-(T) \ge N_-(\Theta^* T \Theta)$.

For all k with $k \le N_-(T) - \beta(G)$, $\lambda_{k+\beta(G)}(T) < 0$, and from the right-hand inequality in (3.21) it follows $\lambda_k(\Theta^* T \Theta) < 0$. Thus $N_-(\Theta^* T \Theta) \ge N_-(T) - \beta(G)$.

Overall, it holds

$$N_-(\Theta^* T \Theta) \le N_-(T) \le N_-(\Theta^* T \Theta) + \beta(G),$$

where we have also shown above that $N_-(T) = \nu(f)$ and $N_-(\Theta^* T \Theta) = \underline{k} - 1$. $\square$

Corollary 3.4.4 (Nodal Domains and Betti Number) *Let $G = (X, E)$ be a connected graph and M a weighted Laplace matrix. Let $k \in \{1, \ldots, |X|\}$ and f a real zero-free eigenfunction of M for $\lambda_k(M)$, then*

$$\underline{k} - \beta(G) \le \mathcal{N}(f) \le \underline{k}.$$

Proof The upper bound follows from the nodal theorem (Theorem 3.3.5), since in our case $\mathcal{N}_0(f) = \mathcal{N}(f)$. From Proposition 3.4.2 and the estimate $\nu(f) \ge \underline{k} - 1$ (Theorem 3.4.3) it follows

$$\mathcal{N}(f) \ge |X| - |E| + \nu(f) \ge |X| - |E| + \underline{k} - 1 \equiv \underline{k} - \beta(G).$$

$\square$

Remark 3.4.5 In the original theorems in [14], it was assumed that $\lambda_k(M)$ is simple. We have adapted the formulation and the proof. It can be shown that almost all weighted Laplace matrices have only simple eigenvalues and all their eigenfunctions are zero-free (i.e., the weighted Laplace matrices that do not meet these conditions form a null set in the space of matrices with the Lebesgue measure). In other words, Theorem 3.4.3 (as well as Theorem 3.5.1 below) holds for the "typical" weighted Laplace matrices.

Exercise 3.4.1 (Fiedler's Theorem on Nodal Domains for Trees) From [51], see also [19, Chapter 4.1.1].

Let $G = (X, E)$ be a <u>tree</u> and M a weighted Laplace matrix for G. Let $k \in \{1, \ldots, |X|\}$ such that there is a zero-free real eigenfunction f of M for the eigenvalue $\lambda_k(M)$. Show:

(a) The eigenvalue $\lambda_k(M)$ is simple.
(b) It holds that $\nu(f) = k - 1$ and $\mathcal{N}(f) = k$.

Hint: One can proceed as in the proof of Theorem 3.4.3. First, one can show that the mapping $\vec{\nabla} : \mathbb{K}^X \to \mathbb{K}^E$ is surjective.

3.5 Nodal Surplus and Flows

Let $G = (X, E)$ be a graph and M a weighted Laplace matrix for G. Let $k \in \{1, \ldots, |X|\}$ such that the eigenvalue $\lambda_k(M)$ is simple. Let $f \in \mathbb{R}^X$ be an eigenfunction of M for $\lambda_k(M)$ and assume that f has no zeros. Define:

$$\text{the } \textbf{nodal surplus} \text{ of } f := \nu(f) - (k - 1).$$

In Theorem 3.4.3 it was shown that the nodal surplus lies between 0 and $\beta(G)$ (in this case $\underline{k} = k$). It is remarkable that there is an explicit formula for the nodal surplus, which was discovered by Berkolaiko [15]. Colin de Verdière [32] has significantly simplified the original proof.[2]

We will first introduce a series of definitions. Let $\vec{E}$ be the set of oriented edges of G,

$$\vec{E} = \{(x, y) \in X \times X : \{x, y\} \in E\}.$$

A **flow** on G is a mapping $\gamma : \vec{E} \to \mathbb{R}$, such that

$$\gamma(x, y) = -\gamma(y, x) \text{ for all } (x, y) \in \vec{E}.$$

We extend every flow γ on G on all of $X \times X$ by zero, which will simplify some expressions. The set

$$\Gamma(G) := \{\gamma : \gamma \text{ is a flow on } G\}$$

with the usual pointwise operations is a real vector space. For a given orientation of G the mapping

$$\Pi : \mathbb{R}^E \to \Gamma(G), \quad (\Pi v)(x, y) = \begin{cases} v(e), & \text{if } x = \iota e, \, y = \tau e, \\ -v(e), & \text{if } y = \iota e, \, x = \tau e, \end{cases} \tag{3.22}$$

is obviously an isomorphism, so $\dim \Gamma(G) = \dim \mathbb{R}^E = |E|$.

Now let M be a weighted Laplace matrix and γ a flow on G. Then we define the matrix $M^\gamma = (M^\gamma_{x,y})_{x,y \in X}$ by

$$M^\gamma_{x,y} := e^{\mathrm{i}\gamma(x,y)} M_{x,y}.$$

---

[2] Note for the instructors: The results of this section are not used in the rest of the text.

The matrix M^γ is Hermitian and defines a self-adjoint linear mapping in $\mathbb{C}^X$. For $\gamma \equiv 0$ we have $M^\gamma = M$.

Theorem 3.5.1 (Berkolaiko: Nodal Surplus Through Flows) *Let $G = (X, E)$ be a connected graph and M a weighted Laplace matrix for G. Let $k \in \{1, \ldots, |X|\}$ such that $\lambda_k(M)$ is a simple eigenvalue of M and the eigenfunctions of M for $\lambda_k(M)$ have no zeros.*

Let $\gamma_1, \ldots, \gamma_m$ be a basis in $\Gamma(G)$, $m := |E|$, and consider the function

$$\Lambda : \mathbb{R}^m \ni t \mapsto \lambda_k(M^{t \cdot \gamma})$$

$$\text{with } t \cdot \gamma := t_1 \gamma_1 + \cdots + t_m \gamma_m \text{ for } t = (t_1, \ldots, t_m). \tag{3.23}$$

(a) There exists an $\varepsilon > 0$ such that

  - *$\Lambda(t)$ is a simple eigenvalue of $M^{t \cdot \gamma}$ for all $t \in (-\varepsilon, \varepsilon)^m$,*
  - *Λ is infinitely differentiable on $(-\varepsilon, \varepsilon)^m$.*

(b) It holds that $\partial_j \Lambda(0) = 0$ for all $j \in \{1, \ldots, m\}$.
(c) Denote

$$\Lambda'' : (-\varepsilon, \varepsilon)^m \ni t \mapsto \left(\partial^2_{jk} \Lambda(t) \right)_{j,k \in \{1,\ldots,m\}} \text{ the Hessian matrix of } \Lambda,$$

then for every real eigenfunction f of M for $\lambda_k(M)$ one has

$$\nu(f) - (k - 1) = \text{the number of negative eigenvalues of } \Lambda''(0).$$

The statement (a) follows from general results about matrices with parameters (Sect. 2.2), and we prove (b) in the following lemma:

Lemma 3.5.2 *For all sufficiently small t, it holds that $\Lambda(-t) = \Lambda(t)$. In particular, $\partial_j \Lambda(0) = 0$ for all $j \in \{1, \ldots, m\}$.*

Proof Let $t = (t_1, \ldots, t_m) \in (-\varepsilon, \varepsilon)^m$ and let f_t be an eigenfunction of $M^{t \cdot \gamma}$ with respect to $\Lambda(t)$. We have $\Lambda(t) f_t = M^{t \cdot \gamma} f_t$. Now we can apply complex conjugation to both sides: For all $x \in X$ we have

$$\Lambda(t) \overline{f_t}(x) = \overline{\Lambda(t) f_t(x)} = \overline{(M^{t \cdot \gamma} f_t)(x)} = M_{x,x} \overline{f_t}(x) + \sum_{y:\, y \sim x} \overline{M_{x,y} e^{it \cdot \gamma(x,y)} f_t(y)}$$

$$= M_{x,x} \overline{f_t}(x) + \sum_{y:\, y \sim x} M_{x,y} e^{-it \cdot \gamma(x,y)} \overline{f_t}(y) = (M^{-t \cdot \gamma} \overline{f_t})(x),$$

so $\Lambda(t)$ is an eigenvalue of $M^{-t\cdot\gamma}$. For small t, $\Lambda(-t)$ is the only eigenvalue of $M^{-t\cdot\gamma}$ in a small open neighborhood of $\lambda_k(M)$, so $\Lambda(t) = \Lambda(-t)$ holds for all sufficiently small t. By differentiating with respect to t_j, we obtain $\partial_j\Lambda(t) = -\partial_j\Lambda(-t)$, and for $t = 0$ it follows that $\partial_j\Lambda(0) = 0$. $\qquad\square$

The rest of the section is dedicated to the proof of (c). This will be broken down into several small steps. We will first perform some calculations in Theorem 3.4.3 in a more precise form.

Lemma 3.5.3 *With the notation of Theorem 3.4.3, the decomposition*

$$\mathbb{K}^E = \operatorname{ran}\vec{\nabla} \oplus \ker\vec{\nabla}^*T \tag{3.24}$$

$$\textit{for} \quad T : \mathbb{K}^E \to \mathbb{K}^E, \quad (Tv)(e) = (-M_e)f(\iota e)f(\tau e)\,v(e),$$

as well as the identity

$$\nu(f) - (k-1) = \max\Big\{ \dim W : W \subset \ker\vec{\nabla}^*T\, subspace$$

$$with\ \langle w, Tw\rangle < 0\, for\ all\ w \in W \setminus \{0\}\Big\}$$

hold. Here, we have $\dim\ker\vec{\nabla}^*T = \beta(G)$.

Proof We continue to use all the objects introduced in the proof of Theorem 3.4.3. By the polar identity for sesquilinear forms, (3.19) implies

$$\big\langle h, \Phi^*(M - \lambda_k I)\Phi g\big\rangle = \langle h, \vec{\nabla}^*T\vec{\nabla}g\rangle \text{ for all } h, g \in \mathbb{K}^X. \tag{3.25}$$

We first show

$$\operatorname{ran}\vec{\nabla} \cap \ker\vec{\nabla}^*T = \{0\}. \tag{3.26}$$

Let $v \in \operatorname{ran}\vec{\nabla} \cap \ker\vec{\nabla}^*T$, then $v = \vec{\nabla}g$ for some $g \in \mathbb{K}^X$ and $0 = \vec{\nabla}^*Tv \equiv \vec{\nabla}^*T\vec{\nabla}g$. From (3.25) it follows that $\big\langle h, \Phi^*(M - \lambda_k I)\Phi g\big\rangle = 0$ for all $h \in \mathbb{K}^X$, so $\Phi^*(M - \lambda_k I)\Phi g = 0$.

Since the mappings $\Phi, \Phi^* : \mathbb{K}^X \to \mathbb{K}^X$ are bijective, we have $(M - \lambda_k I)\Phi g = 0$ and thus $\Phi g \in \ker(M - \lambda_k I) = \mathbb{K}f$. So there is a $c \in \mathbb{K}$ with $\Phi g \equiv fg = cf$. Since f has no zeros, we have $g = c\mathbb{1}$, so $v = \vec{\nabla}g = 0$. This proves (3.26).

Now we prove the representation (3.24). Due to (3.26), we can simply count the dimensions. Since $T : \mathbb{K}^E \to \mathbb{K}^E$ is bijective, we have

$$\ker\vec{\nabla}^*T = \big\{v \in \mathbb{K}^E : \langle\vec{\nabla}^*Tv, g\rangle = 0 \text{ for all } g \in \mathbb{K}^X\big\}$$

$$= \big\{v \in \mathbb{K}^E : \langle v, T\vec{\nabla}g\rangle = 0 \text{ for all } g \in \mathbb{K}^X\big\} = \operatorname{ran}(T\vec{\nabla})^{\perp},$$

$$\dim \ker \vec{\nabla}^* T = \dim \operatorname{ran}(T\vec{\nabla})^{\perp}$$

$$= \dim \mathbb{K}^E - \dim \operatorname{ran} T\vec{\nabla} = \dim \mathbb{K}^E - \dim \operatorname{ran} \vec{\nabla} \overset{(3.20)}{=} \beta(G),$$

and thus $\dim \operatorname{ran} \vec{\nabla} + \dim \ker \vec{\nabla}^* T = \dim \mathbb{K}^E$, which completes the proof of the decomposition (3.24). We mention an additional useful property of this decomposition: If $v \in \operatorname{ran} \vec{\nabla}$ and $w \in \ker \vec{\nabla}^* T$, then $v = \vec{\nabla} g$ for some $g \in \mathbb{K}^X$ and

$$\langle v, Tw \rangle = \langle \vec{\nabla} g, Tw \rangle = \langle g, \underbrace{\vec{\nabla}^* Tw}_{=0} \rangle = 0. \tag{3.27}$$

Let $U \subset \mathbb{K}^E$ be a subspace. Thanks to (3.24), we have $U = V \oplus W$ with uniquely defined subspaces $V \subset \operatorname{ran} \vec{\nabla}$ and $W \subset \ker \vec{\nabla}^* T$. If $u \in U$ with $u = v + w$ and $v \in V, w \in W$, then $\langle u, Tu \rangle = \langle v, Tv \rangle + \langle w, Tw \rangle$ thanks to (3.27). So $\langle u, Tu \rangle < 0$ holds for all $u \in U \setminus \{0\}$ if and only if

$$\langle v, Tv \rangle < 0 \text{ for all } v \in V \setminus \{0\} \quad \underline{and} \quad \langle w, Tw \rangle < 0 \text{ for all } w \in W \setminus \{0\}.$$

It follows

$$\nu(f) = N_-(T) = \max \left\{ \dim U : \quad U \subset \mathbb{K}^E \text{ with } \langle u, Tu \rangle < 0 \text{ for all } u \in U \setminus \{0\} \right\}$$

$$= \max \Big\{ \dim U : \quad U = V \oplus W,$$

$$V \subset \operatorname{ran} \vec{\nabla} \text{ with } \langle v, Tv \rangle < 0 \text{ for all } v \in V \setminus \{0\},$$

$$W \subset \ker \vec{\nabla}^* T \text{ with } \langle w, Tw \rangle < 0 \text{ for all } w \in W \setminus \{0\} \Big\}$$

$$= N_V + N_W,$$

where

$$N_V := \max \left\{ \dim V : V \subset \operatorname{ran} \vec{\nabla} \text{ with } \langle v, Tv \rangle < 0 \text{ for all } v \in V \setminus \{0\} \right\}$$

$$\equiv N_-(\Theta^* T \Theta) = k - 1,$$

$$N_W := \max \left\{ \dim W : W \subset \ker \vec{\nabla}^* T \text{ with } \langle w, Tw \rangle < 0 \text{ for all } w \in W \setminus \{0\} \right\},$$

so $\nu(f) = (k - 1) + N_W$. $\qquad \square$

Now we would like to take a closer look at the Hessian matrix Λ'' of the function Λ in (3.23).

Lemma 3.5.4 *The number of negative eigenvalues of $\Lambda''(0)$ is independent of the choice of the basis $\gamma_1, \ldots, \gamma_m$.*

Proof Let $\widehat{\gamma}_1, \ldots, \widehat{\gamma}_m$ be another basis of $\Gamma(G)$ and

$$\widehat{\Lambda} : \; \mathbb{R}^m \ni t \mapsto \lambda_k(M^{t \cdot \widehat{\gamma}}), \quad t \cdot \widehat{\gamma} := t_1 \widehat{\gamma}_1 + \cdots + t_m \widehat{\gamma}_m \text{ for } t = (t_1, \ldots, t_m).$$

There is a real invertible $m \times m$ transition matrix $B = (b_{i,j})$, such that

$$\widehat{\gamma}_j = \sum_{i=1}^{m} b_{i,j} \gamma_i \text{ for all } j \in \{1, \ldots, m\}.$$

So, with $t = (t_1, \ldots, t_m)$,

$$t \cdot \widehat{\gamma} = \sum_{j=1}^{m} t_j \gamma_j = \sum_{j=1}^{m} t_j \sum_{i=1}^{m} b_{i,j} \gamma_i = \sum_{i=1}^{m} \Big(\sum_{j=1}^{m} b_{i,j} t_j \Big) \gamma_i = \sum_{i=1}^{m} (Bt)_i \gamma_i = (Bt) \cdot \gamma$$

and thus $\widehat{\Lambda}(t) = \Lambda(Bt)$. Therefore

$$\partial_j \widehat{\Lambda}(t) = \partial_{t_j} \Lambda(Bt) = \sum_{r=1}^{m} \partial_r \Lambda(Bt) b_{r,j},$$

$$\partial_{ij}^2 \widehat{\Lambda}(t) = \partial_i (\partial_j \widehat{\Lambda})(t) = \sum_{r,s=1}^{m} \partial_{sr}^2 \Lambda(Bt) b_{s,i} b_{r,j} = \big(B^* \Lambda''(Bt) B \big)_{i,j},$$

in particular $\widehat{\Lambda}''(0) = B^* \Lambda''(0) B$. Thus, $\Lambda''(0)$ and $\widehat{\Lambda}''(0)$ have the same number of negative eigenvalues according to Sylvester's law of inertia (Corollary 2.1.11). $\square$

Now we would like to find a "good" basis in $\Gamma(G)$ such that $\Lambda''(0)$ has the simplest possible form. To recall, the mapping $\Pi : \mathbb{R}^E \to \Gamma(G)$ was defined in (3.22).

Lemma 3.5.5 Let $\gamma, \widehat{\gamma} \in \Gamma(G)$ with $\widehat{\gamma} \in \Pi(\mathrm{ran}\, \vec{\nabla})$, then $\sigma(M^{\gamma+\widehat{\gamma}}) = \sigma(M^{\gamma})$.

Proof The condition $\widehat{\gamma} \in \Pi(\mathrm{ran}\, \vec{\nabla})$ means that for some $h \in \mathbb{R}^X$ we have

$$\widehat{\gamma}(x, y) = h(x) - h(y) \text{ for all } (x, y) \in \vec{E}.$$

For all $f \in \mathbb{K}^X$ and $x \in X$ we have

$$(M^{\gamma+\widehat{\gamma}} f)(x) = M_{x,x} f(x) + \sum_{y:\, y \sim x} M_{x,y} e^{\mathrm{i}\gamma(x,y) + \mathrm{i}\widehat{\gamma}(x,y)} f(y)$$

$$= M_{x,x} f(x) + \sum_{y:\, y \sim x} M_{x,y} e^{\mathrm{i}\gamma(x,y)} e^{\mathrm{i}(h(x) - h(y))} f(y)$$

$$= e^{ih(x)}\left(M_{x,x}e^{-ih(x)}f(x) + \sum_{y:\,y\sim x} M_{x,y}e^{i\gamma(x,y)}e^{-ih(y)}f(y)\right)$$

$$= (\Theta^{-1}M^{\gamma}\Theta f)(x)$$

for the bijective mapping $\Theta:\ \mathbb{K}^X \to \mathbb{K}^X$ with $(\Theta f)(x) := e^{-ih(x)}f(x)$.

Thus $M^{\gamma+\widehat{\gamma}} = \Theta^{-1}M^{\gamma}\Theta$ and the claim follows from Proposition 1.4.6. $\qquad\square$

Since Π is an isomorphism, the decomposition (3.24) also leads to the decomposition

$$\Gamma(G) \equiv \Pi(\mathbb{R}^E) = \Pi(\operatorname{ran}\vec{\nabla}) \oplus \Pi(\ker\vec{\nabla}^*T). \tag{3.28}$$

Let

$$n := |X|, \quad \beta := \beta(G) \equiv m - n + 1$$

and choose:

- a basis $\widehat{\gamma}_1, \ldots, \widehat{\gamma}_{n-1}$ in $\Pi(\operatorname{ran}\vec{\nabla})$,
- a basis $\gamma_1, \ldots, \gamma_{\beta}$ in $\Pi\left(\ker\vec{\nabla}^*T\right) \subset \mathbb{R}^E$.

Thanks to the decomposition (3.28), the flows $\widehat{\gamma}_1, \ldots, \widehat{\gamma}_{n-1}$, $\gamma_1, \ldots, \gamma_{\beta}$ form a basis in $\Gamma(G)$. Consider the associated function Λ. For arbitrary $t_1, \ldots, t_{n-1}, s_1, \ldots, s_{\beta} \in \mathbb{R}$, due to Lemma 3.5.5, we have

$$\Lambda(t_1, \ldots, t_{n-1}, s_1, \ldots, s_{\beta}) = \lambda_k(M^{t_1\widehat{\gamma}_1+\cdots+t_{n-1}\widehat{\gamma}_{n-1}+s_1\gamma_1+\cdots+s_{\beta}\gamma_{\beta}})$$

$$= \lambda_k(M^{s_1\gamma_1+\cdots+s_{\beta}\gamma_{\beta}}) =: \lambda(s_1, \ldots, s_{\beta}),$$

and the Hessian matrix $\Lambda''(0)$ has the block form

$$\Lambda''(0) = \begin{pmatrix} 0 & 0 \\ 0 & \lambda''(0) \end{pmatrix},$$

where λ'' is the Hessian matrix of λ. If $N_-(B)$ stands for the number of negative eigenvalues of a linear mapping B, we have

$$N_-\left(\Lambda''(0)\right) = N_-\left(\lambda''(0)\right) = \max\left\{\dim S:\ S \subset \mathbb{R}^{\beta}\ \text{subspace}\right.$$

$$\left. \text{with } \sum_{i,j=1}^{\beta} \partial_{ij}^2\lambda(0)s_is_j < 0 \text{ for all } (s_1, \ldots, s_{\beta}) \in S \setminus \{0\}\right\}. \tag{3.29}$$

Now we use the identity

$$\sum_{i,j=1}^{\beta} \partial_{ij}^2 \lambda(0) s_i s_j = \frac{\mathrm{d}^2}{\mathrm{d}t^2}\Big|_{t=0} \lambda\big(t(s_1,\ldots,s_\beta)\big) = \mu''_{s_1\gamma_1+\cdots+s_\beta\gamma_\beta}(0),$$

$$\text{where}\quad \mu_\gamma : \mathbb{R} \ni t \mapsto \lambda_k(M^{t\gamma}) \quad \text{for } \gamma \in \Gamma(G).$$

From (3.29) it follows that

$$\begin{aligned}
N_-\big(\Lambda''(0)\big) &= \max\Big\{ \dim S : \ S \subset \mathbb{R}^\beta \text{ with } \mu''_{s_1\gamma_1+\cdots+s_\beta\gamma_\beta}(0) < 0 \\[4pt]
&\qquad\qquad \text{for all } (s_1,\ldots,s_\beta) \in S \setminus \{0\}\Big\} \\[6pt]
&\equiv \max\Big\{ \dim W : \ W \subset \ker \vec{\nabla}^* T \subset \mathbb{R}^E \text{ with } \mu''_{\Pi w}(0) < 0 \\[4pt]
&\qquad\qquad \text{for all } w \in W \setminus \{0\}\Big\}.
\end{aligned} \tag{3.30}$$

Lemma 3.5.6 *For all $w \in \ker \vec{\nabla}^* T \subset \mathbb{R}^E$, we have $\mu''_{\Pi w}(0) = 2\langle w, Tw\rangle$.*

Proof Let $\gamma \in \Gamma(G)$. For $t \in \mathbb{R}$, let

$$\mu(t) := \mu_\gamma(t), \quad M(t) := M^{t\gamma}$$

and let $f(t)$ be an eigenfunction of $M(t)$ to the eigenvalue $\mu(t)$ with $\|f(t)\| = 1$ and $f(0) = f$, so that for a sufficiently small $\varepsilon > 0$ the function $(-\varepsilon, \varepsilon) \ni t \mapsto f(t) \in \mathbb{C}^X$ is smooth. This choice is possible according to Theorem 2.2.2.

For all $t \in (-\varepsilon, \varepsilon)$, we have $\big[M(t) - \mu(t)I\big]f(t) = 0$. Using the Leibniz rule for derivatives, we differentiate both sides twice with respect to t:

$$\big[M'(t) - \mu'(t)I\big]f(t) + \big[M(t) - \mu(t)I\big]f'(t) = 0,$$

$$\big[M''(t) - \mu''(t)I\big]f(t) + 2\big[M'(t) - \mu'(t)I\big]f'(t) + \big[M(t) - \mu(t)I\big]f''(t) = 0.$$

The scalar product of the last line with $f(t)$ yields

$$\Big\langle f(t), M''(t)f(t)\Big\rangle - \mu''(t) + 2\Big\langle f(t), \big[M'(t) - \mu'(t)I\big]f'(t)\Big\rangle$$

$$+ \Big\langle f(t), \big[M(t) - \mu(t)I\big]f''(t)\Big\rangle = 0.$$

Since the matrix $M(t)$ is Hermitian for each t, all its derivatives with respect to t are also Hermitian. It follows that the last term on the left-hand side is also zero:

$$\left\langle f(t), \left[M(t) - \mu(t)I\right]f''(t)\right\rangle = \left\langle \underbrace{\left[M(t) - \mu(t)I\right]f(t)}_{=0}, f''(t)\right\rangle = 0,$$

so we get

$$\mu''(t) = \left\langle f(t), M''(t)f(t)\right\rangle + 2\left\langle f(t), \left[M'(t) - \mu'(t)I\right]f'(t)\right\rangle.$$

Write $\gamma = s_1\gamma_1 + \cdots + s_m\gamma_m$ (where $\gamma_1, \ldots, \gamma_m$ is a basis in $\Gamma(G)$ and $s_1, \ldots, s_m \in \mathbb{K}$), then

$$\mu(t) = \Lambda(ts_1, \ldots, ts_m) \equiv \Lambda(ts), \quad \mu'(t) = s_1\partial_1\Lambda(ts) + \cdots + s_m\partial_m\Lambda(ts).$$

Thanks to Lemma 3.5.2, we have $\partial_j\Lambda(0) = 0$ for all j, and thus we obtain $\mu'(0) = 0$. Therefore,

$$\begin{aligned}
\mu''(0) &= \langle f, M''(0)f\rangle + 2\langle f, M'(0)f'(0)\rangle \\
&= \langle f, M''(0)f\rangle + 2\langle M'(0)f, f'(0)\rangle.
\end{aligned} \tag{3.31}$$

We have $M(t) = \left(e^{it\gamma(x,y)}M_{x,y}\right)_{x,y\in X}$ and it follows

$$M'(0) = \left(i\gamma(x, y)M_{x,y}\right)_{x,y\in X}, \qquad M''(0) = \left(-\gamma(x, y)^2 M_{x,y}\right)_{x,y\in X}.$$

Therefore, we have

$$\left[M'(0)f\right](x) = i\sum_{y:\, y\sim x}\gamma(x, y)M_{x,y}f(y), \quad x \in X. \tag{3.32}$$

Now we use the above calculations for $\gamma = \Pi w$ with $w \in \ker \vec{\nabla}^* T$, i.e.,

$$\gamma(x, y) = \begin{cases} w(e), & x = \iota e \text{ and } y = \tau e, \\ -w(e), & x = \tau e \text{ and } y = \iota e, \\ 0, & \text{otherwise.} \end{cases} \tag{3.33}$$

We first find an explicit formula for $\vec{\nabla}^* T$. Let $h \in \mathbb{K}^X$ and $u \in \mathbb{K}^E$, then

$$\langle h, \vec{\nabla}^* Tu\rangle = \langle\vec{\nabla}h, Tu\rangle = \sum_{e\in E}\left(\overline{h(\iota e)} - \overline{h(\tau e)}\right)t(e)u(e)$$

$$= \sum_{x\in X}\overline{h(x)}\left(\sum_{e:\,\iota e=x}t(e)u(e) - \sum_{e:\,\tau e=x}t(e)u(e)\right),$$

thus

$$(\vec{\nabla}^*Tu)(x) = \sum_{e:\iota e=x} t(e)u(e) - \sum_{e:\tau e=x} t(e)u(e), \quad u \in \mathbb{K}^E, \quad x \in X,$$

and we have $t(e) = -M_e f(\iota e) f(\tau e)$. Then the formula (3.32) can be rewritten:

$$\begin{aligned}
[M'(0)f](x) &= \mathrm{i}\sum_{e:\iota e=x} w(e)M_e f(\tau e) - \mathrm{i}\sum_{e:\tau e=x} w(e)M_e f(\iota e)\\[2mm]
&= -\left(\mathrm{i}\sum_{e:\iota e=x} \frac{t(e)w(e)}{f(\iota e)} - \mathrm{i}\sum_{e:\tau e=x} \frac{t(e)w(e)}{f(\tau e)}\right)\\[2mm]
&= -\frac{\mathrm{i}}{f(x)}\left(\vec{\nabla}^*Tw\right)(x) = 0
\end{aligned}$$

From (3.31) it follows that

$$\mu''(0) = \langle f, M''(0)f\rangle = -\sum_{x,y\in X} M_{x,y}\gamma(x,y)^2 f(x)f(y),$$

and with the help of (3.33) we obtain

$$\begin{aligned}
\mu''(0) &= -\sum_{e\in E} M_{\iota e,\tau e}\gamma(\iota e,\tau e)^2 f(\iota e)f(\tau e) - \sum_{e\in E} M_{\tau e,\iota e}\gamma(\tau e,\iota e)^2 f(\tau e)f(\iota e)\\[2mm]
&= -2\sum_{e\in E} M_e f(\iota e)f(\tau e)w(e)^2 \equiv 2\langle w, Tw\rangle.
\end{aligned}$$

$\square$

It remains to summarize all calculations:

End of the Proof of Theorem 3.5.1 (Nodal Surplus Through Flows)

$$\begin{aligned}
N_-\left(\Lambda''(0)\right) &= \max\Big\{ \dim W : \ W \subset \ker \vec{\nabla}^*T \subset \mathbb{R}^E \ \text{subspace}\\[2mm]
&\qquad\quad \text{with } \mu''_{\Pi w}(0) < 0 \text{ for all } w \in W \setminus \{0\}\Big\} \qquad \text{(holds by (3.30))}\\[3mm]
&= \max\Big\{ \dim W : \ W \subset \ker \vec{\nabla}^*T \subset \mathbb{R}^E \ \text{subspace}\\[2mm]
&\qquad\quad \text{with } \langle w, Tw\rangle < 0 \text{ for all } w \in W \setminus \{0\}\Big\} \qquad \text{(Lemma 3.5.6)}\\[3mm]
&= \nu(f) - (k-1). \qquad\qquad\qquad\qquad\qquad\qquad\qquad\quad \text{(Lemma 3.5.3)}
\end{aligned}$$

$\square$

With Theorem 3.5.1 we can derive a new characterization of trees that was first found by Band [10].

Theorem 3.5.7 (Band: Nodal Surplus and Trees) *Let G be a connected graph and M a weighted Laplace matrix for G such that all eigenvalues of M are simple and the eigenfunctions have no zeros. Then the following statements are equivalent:*

(a) *The nodal surpluses of all eigenfunctions of M are zero, i.e., real eigenfunctions of M to $\lambda_k(M)$ have for each $k \in \{1, \ldots, |X|\}$ exactly $k - 1$ sign-changing edges.*

(b) *G is a tree.*

Proof The direction (b) $\Rightarrow$(a) follows from Theorem 3.4.3, since the first Betti number of a tree is always zero.

Assume now that (a) is fulfilled and that $\beta(G) \geq 1$. We denote $n := |X|$ and $m := |E|$. For each $k \in \{1, \ldots, n\}$ we choose a real eigenfunction f_k of M to $\lambda_k(M)$ and consider the following linear mappings $T_k : \mathbb{R}^E \to \mathbb{R}^E$:

$$(T_k w)(e) := t_k(e) w(e) \text{ with } t_k(e) := -M_e f_k(\iota e) f_k(\tau e).$$

Because $\dim \ker \vec{\nabla}^* T_1 = \beta(G) \geq 1$ we find some $w \in \ker \vec{\nabla}^* T_1 \subset \mathbb{R}^E$ with $w \neq 0$. Consider the functions

$$\mu_k : \mathbb{R} \ni s \mapsto \lambda_k(M^{s\Pi w}), \quad k \in \{1, \ldots, n\}.$$

Let $\gamma_1, \ldots, \gamma_m$ be a basis in $\Gamma(G)$ and

$$\Lambda_k : \mathbb{R}^m \ni (t_1, \ldots, t_m) \mapsto \lambda_k(M^{t \cdot \gamma}).$$

By assumption, $\nu(f_k) = k - 1$, so $\Lambda_k''(0)$ has no negative eigenvalues according to theorem 3.5.1(c), and $\langle t, \Lambda''(0)t \rangle \geq 0$ for all $t \in \mathbb{R}^m$ by the min-max principle. Let $t = (t_1, \ldots, t_m) \in \mathbb{R}^m$ with $\Pi w = t_1 \gamma_1 + \cdots + t_m \gamma_m \equiv t \cdot \gamma$, then

$$\mu_k''(0) = \frac{d^2}{ds^2}\Big|_{s=0} \lambda_k(M^{st \cdot \gamma}) = \frac{d^2}{ds^2}\Big|_{s=0} \Lambda_k(st) = \langle t, \Lambda''(0)t \rangle \geq 0. \tag{3.34}$$

The previous inequality (3.34) holds for all $k \in \{1, \ldots, n\}$, but for $k = 1$ it can be strengthened. The eigenfunction f_1 for the smallest eigenvalue of M has a constant sign (and no zeros), so $t_1(e) > 0$ for all $e \in E$. By Lemma 3.5.6 we have

$$\mu_1''(0) = 2 \langle w, T_1 w \rangle = 2 \sum_{e \in E} t_1(e) w(e)^2 > 0.$$

For the function

$$C : \mathbb{R} \ni s \mapsto \sum_{k=1}^{n} \mu_k(s) \in \mathbb{R}$$

we then obtain $C''(0) > 0$.

We note that the diagonal elements of M^γ are independent of γ. For each $s \in \mathbb{R}$ the trace theorem for $M^{s\Pi w}$ holds:

$$C(s) = \operatorname{tr} M^{s\Pi w} \equiv \sum_{x \in X} (M^{s\Pi w})_{x,x} \equiv \sum_{x \in X} M_{x,x}.$$

Since the right-hand side is independent of s, it follows that $C''(0) = 0$: contradiction.

Therefore, $\beta(G) = 0$ must hold and thus G is a tree. $\qquad\qquad\square$

Remark 3.5.8 For the weighted Laplace matrices, it was always assumed that $M_{x,y} < 0$ for $x \sim y$. This condition can be weakened: one can only require that $x \sim y$ (for $x \neq y$) is equivalent to $M_{x,y} \neq 0$. Such matrices M are often called generalized weighted Laplace matrices. In this case, the sign-changing edges can be redefined: we say that e is an M-sign-changing edge for a function $f \in \mathbb{R}^X \setminus \{0\}$ if $(-M_e) f(\iota e) f(\tau e) < 0$. The number of such edges is denoted by $\nu_M(f)$. A small adjustment of the above proofs shows that the two Theorems 3.4.3 and 3.5.1 also hold for generalized weighted Laplace matrices M with $\nu_M(f)$ instead of $\nu(f)$. In addition, one can also prove some statements about eigenfunctions with zeros, see the articles [5, 119] and the book [19, Chapter 3]. For this new class of graph matrices M, however, we are not aware of a direct connection between $\nu_M(f)$ and the number of nodal domains (for the nodal Theorem 3.3.5 the signs of M_e are really important). The nodal domains of eigenfunctions for generalized weighted Laplace matrices are also relevant in the cluster analysis of graphs [53, 71].

Planarity and Colin de Verdière Invariant 4

A graph is planar if it can be drawn in the plane without self-intersections. In this chapter, we will learn how to identify planar graphs and what additional properties they and their graph matrices possess.

4.1 Euler's Formula and *N*-Color Theorems

The concept of planarity was introduced in Sect. 1.1: A graph is called planar if a planar drawing (i.e., a crossing-free drawing in the plane) can be found for it. For the study of planar graphs, we will use subdivisions, edge contractions, and the formation of minors (Definition 1.1.25). Our presentation is very compact and we refer to the books by Aigner [2], Diestel [45], and West [113] for further study.

Let us gather some initial ideas regarding planar graphs:

- A graph is planar if and only if all its connected components are planar.
- A graph is planar if and only if all its subdivisions are planar. In fact, the same drawing can be used for the graph and its subdivisions.
- If a graph is planar, it remains planar after removing vertices and edges.
- If a graph G is planar, it remains planar after the contraction of any edge $e = \{x, y\}$. This can be seen as follows: First consider any planar drawing of G. The new vertex x_e, which is created when the edge e is contracted, is placed in the middle of the edge e. Then all vertices of $G \cdot e$, which were previously connected with x or y, can also be connected without crossings to x_e by drawing the corresponding edges very close to the old edges of G (Fig. 4.1).
- It follows that all minors of a planar graph are also planar.

Let a planar drawing Γ of a planar graph G be given. A **face** in this drawing is a connected component of $\mathbb{R}^2 \setminus \Gamma$ (Fig. 4.2). Since Γ is always compact, there

© The Author(s), under exclusive license to Springer Nature Switzerland AG 2025   149
K. Naderi, K. Pankrashkin, *Introduction to Spectral Graph Theory*, Compact
Textbooks in Mathematics, https://doi.org/10.1007/978-3-032-01708-6_4

Fig. 4.1 Edge contraction for planar graphs

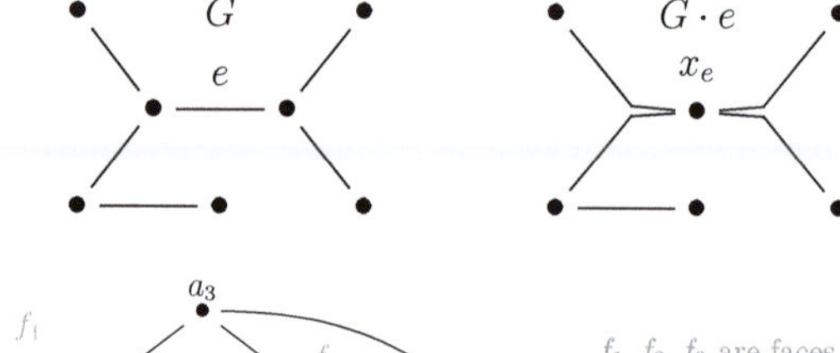

Fig. 4.2 A planar drawing and the associated faces

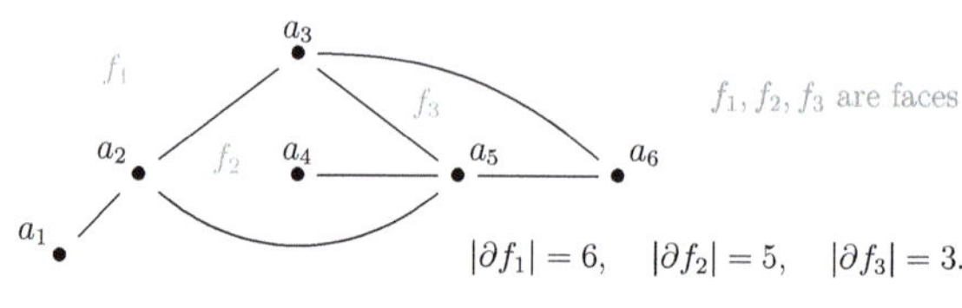

is exactly one face for each drawing that is unbounded. We make the following remarks about the faces:

- Given a planar drawing of a graph G with connected components $G_1, \ldots, G_n$, the drawing Γ_j of each G_j is a connected set. For each $i \in \{1, \ldots, n\} \setminus \{j\}$, Γ_i lies in a face of Γ_j.
- If an isolated vertex or a pendant vertex with its associated edge is deleted in a planar drawing of a graph, the number of faces remains unchanged.
- If a new isolated vertex is drawn in a planar drawing of a graph, or if one draws a new edge between two different connected components that does not intersect any of the existing edges, then the number of faces remains unchanged.

From these considerations, it follows inductively that there is exactly one face in every planar drawing of a forest.

An edge e is called a **boundary edge** of a face f, if the drawing of e lies in the closure of f. The same applies to vertices lying in the closure of f, which are then referred to as **boundary vertices** of f. The **boundary** ∂f of a face f is the subgraph of G consisting of all boundary edges and vertices of f (in this case, it is said that f is bordered by ∂f). By construction, f is also a face in the given drawing of ∂f. For each boundary edge $e \in \partial f$ there are two cases:

A. f lies on both sides of e.
B. Two different faces lie on the two sides of e.

Note that when removing the edges of type A, the number of faces remains unchanged. The **length of the boundary** $|\partial f|$ of the face f is defined by

$$|\partial f| := 2 \cdot \text{(the number of boundary edges of type A)}$$

$$+ \text{(the number of boundary edges of type B)}.$$

For a class of graphs, the boundaries of the faces have a simple structure:

Lemma 4.1.1 *In every planar drawing of a 2-connected planar graph, each face is bordered by a cycle.*

Proof Let Γ be a planar drawing of a 2-connected graph G and let f be a face in this drawing with the boundary ∂f.

Assume that ∂f contains no cycle, then ∂f is a forest, and the drawing of ∂f has only one face (which is then unbounded). Since f is also a face of ∂f, f must be the unbounded face of Γ. Since Γ is contained in the drawing of $\mathbb{R}^2 \setminus f$, Γ coincides with the drawing of ∂f, and it follows $G = \partial f$, so G is a forest and, therefore, not a 2-connected graph: contradiction. Therefore, ∂f must contain a cycle.

Let C be a cycle in ∂f. Since each side of the drawing of C is connected, the face f lies on one side of C. Assume that C is not a connected component of ∂f, then there is a vertex x of ∂f that does not lie on C, but is connected to a vertex y on C by an edge. There is no path between x and another vertex on C, as such a path together with the edge $\{x, y\}$ would cut off a connected component within f. Then there is no path between x and $C \setminus x$ in $G \setminus y$, so G is not 2-connected: contradiction. Therefore, C must be a connected component of ∂f. If ∂f has another connected component C', there is no path between C and C' in G, so G is not connected, in particular not 2-connected: contradiction. Therefore, $\partial f = C$. $\square$

Theorem 4.1.2 (Euler's Formula) *Let $G = (X, E)$ be a connected planar graph. Choose a planar drawing of G and let*

$$F := \text{the set of faces,}$$

then

$$|X| + |F| - |E| = 2. \tag{4.1}$$

Proof Let $G' = (X, E')$ be a spanning tree of G. The given planar drawing of G induces a drawing of G' by keeping only the edges of G'. In the drawing of G', there are $|X|$ vertices, $|E'| \equiv |X| - 1$ edges, and one face, so the Euler formula for G' is satisfied. Now we sequentially draw the remaining $|X| - |E| + 1$ edges of G. In each step, the set of vertices remains unchanged, but an additional edge and an additional face are obtained, so the identity (4.1) remains valid. $\square$

Theorem 4.1.3 (Upper Bound for the Number of Edges in Planar Graphs) *Let $G = (X, E)$ be a planar graph with at least 3 vertices. Then $|E| \leq 3|X| - 6$. If G does not contain any triangles, then $|E| \leq 2|X| - 4$.*

Proof W.l.o.g, we can assume that the graph is connected (otherwise consider each connected component individually).

Consider a planar drawing Γ of G and let F be the set of faces. From the definition of the length of the boundary it follows

$$2|E| = \sum_{f \in F} |\partial f|. \tag{4.2}$$

We first prove the following auxiliary statement:

(H) If for a face $f \in F$ the boundary ∂f does not contain a cycle, then $|\partial f| \geq 4$.

Note that f is also a face for the induced drawing Γ' of ∂f. Since in this case ∂f is a forest, simply $f = \mathbb{R}^2 \setminus \Gamma'$. From $\Gamma \subset (\mathbb{R}^2 \setminus f) = \Gamma'$ it follows $\Gamma = \Gamma'$ and thus $G = \partial f$, i.e., G is also a forest and each edge of G is a boundary edge for f. Since G is also connected, G is a tree, so $|E| = |X| - 1 \geq 2$. Since the same face f lies on each side of each edge, $|\partial f| = 2|E| \geq 4$. This proves (H).

If there were $f \in F$ with $|\partial f| \leq 2$, then ∂f would contain at most two edges. Since there are at least three edges in each cycle, ∂f is cycle-free, so $|\partial f| \geq 4$ by (H): contradiction. Therefore, $|\partial f| \geq 3$ for all $f \in F$, and from (4.2) it follows

$$2|E| = \sum_{f \in F} |\partial f| \geq \sum_{f \in F} 3 = 3|F|, \quad \text{that is, } |F| \leq \frac{2}{3}|E|.$$

From the Euler formula it follows

$$|E| = |X| + |F| - 2 \leq |X| + \frac{2}{3}|E| - 2,$$

and thus $|E| \leq 3|X| - 6$.

Now assume that G does not contain any triangles. Let $f \in F$. If ∂f is cycle-free, then $|\partial f| \geq 4$ by (H). Otherwise, let m be the minimal length of a cycle in ∂f. The case $m = 3$ is impossible (a cycle with 3 vertices is a triangle, which is excluded by assumption), so $m \geq 4$, then $|\partial f| \geq m \geq 4$. Thus, we have shown $|\partial f| \geq 4$ for all $f \in F$. From (4.2) we get

$$2|E| = \sum_{f \in F} |\partial f| \geq \sum_{f \in F} 4 = 4|F|, \quad \text{that is, } |F| \leq \frac{1}{2}|E|,$$

and from the Euler formula it follows that

$$|E| = |X| + |F| - 2 \leq |X| + \frac{1}{2}|E| - 2,$$

thus $|E| \leq 2|X| - 4$. $\qquad\square$

Although the previous considerations are very elementary, they can be used to prove useful statements.

Theorem 4.1.4 (Kempe–Heawood: Five-Color Theorem) *For every planar graph $G = (X, E)$ one has $\chi(G) \leq 5$.*

Proof We use induction on $|X|$. For $|X| \leq 5$, the statement is trivial (one can give all vertices different colors).

Now let $|X| \geq 6$, so that for all planar graphs $G' = (X', E')$ with $|X'| \leq |X| - 1$, the inequality $\chi(G') \leq 5$ holds. Consider a planar drawing of G. The inequality $\deg x \geq 6$ for all $x \in X$ is impossible: One would obtain a contradiction with Theorem 4.1.3 through

$$6|X| \leq \sum_{x \in X} \deg x = 2|E| \leq 2(3|X| - 6) = 6|X| - 12.$$

Therefore, there is a vertex $a \in X$ with $\deg a \leq 5$. The graph

$$G' := G \setminus a$$

is planar and has $|X| - 1$ vertices, so by the induction hypothesis there is a vertex coloring $c : X \setminus \{a\} \to \{1, 2, 3, 4, 5\}$ for G'.

(i) If the neighbors of a carry at most four of the possible five colors (this is particularly the case for $\deg a \leq 4$), there is a free color for a: Thus, we have constructed a vertex coloring for G with five colors.

(ii) Only the following case remains:

$$\deg a = 5, \qquad \text{the neighbors of } a \text{ all have different colors.}$$

Let $b_1, \ldots, b_5$ be the neighbors of a in cyclic order and we can assume w.l.o.g. $c(b_i) = i$. For $i, j \in \{1, \ldots, 5\}$ with $i \neq j$ consider the sets

$$X_{i,j} = \{x \in X \setminus \{a\} : c(x) \in \{i, j\}\}$$

and let $G'_{i,j} := G_{X_{i,j}}$ be the corresponding induced graphs in G.

(iia) If b_i and b_j lie in two different connected components of $G'_{i,j}$, one can swap the colors i and j in one component. Then b_i and b_j receive the same color and $b_1, \ldots, b_5$ occupy only four colors, so case (i) applies and thus $\chi(G) \leq 5$.

(iib) The following case remains: For all $i \neq j$ there is a path $P_{i,j}$ from b_i to b_j in $G'_{i,j}$. From $P_{1,3}$ and the edges $\{a, b_1\}$ and $\{a, b_3\}$ a cycle in G is formed. By construction, b_2 lies inside the cycle and b_4 outside the cycle and by the Jordan curve theorem there must be an intersection between $P_{2,4}$ and $P_{1,3}$. Since the paths run in G' and the drawing is crossing-free, the only possible intersection is at a vertex $x \notin \{a\}$ (see Fig. 4.3).

Fig. 4.3 The paths $P_{1,3}$ and $P_{2,4}$

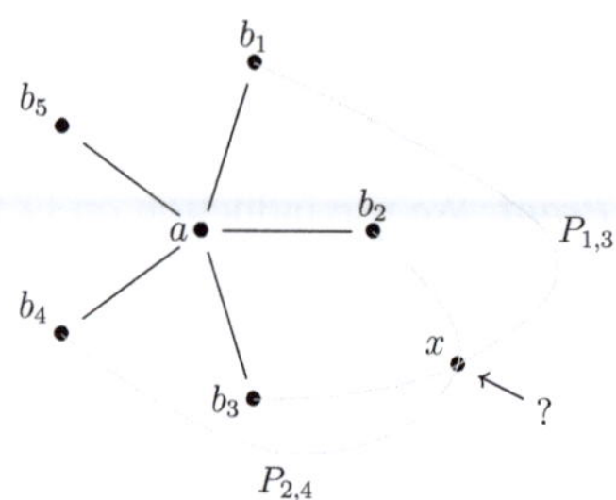

Since $x \in P_{1,3}$, we have $c(x) \in \{1, 3\}$, and since $x \in P_{2,4}$ it must hold that $c(x) \in \{2, 4\}$. The two conditions cannot be fulfilled simultaneously, so case (iib) is impossible.   □

Actually, a stronger theorem holds, which we state here without proof:

Theorem 4.1.5 (Appel–Haken: Four-Color Theorem) *For every planar graph G, $\chi(G) \leq 4$ holds.*

Remark 4.1.6 The Four-Color Theorem was announced in 1879 by Alfred Kempe. In 1890, Percy John Heawood found an error in the proof, but he was able to prove the Five-Color Theorem using Kempe's proof idea. The first complete proof of the four-color theorem was only found in 1976 by Kenneth Appel and Wolfgang Haken [8,9]. As part of the proof, 1936 special graphs had to be analyzed with a computer program. Later, further proof ideas were proposed, but in every existing proof at least 633 graphs must be individually examined. The Four-Color Theorem is the most famous mathematical theorem for which no direct proof without computer assistance has been found so far.

4.2   Planarity Criteria

From Theorem 4.1.3 it also follows:

Corollary 4.2.1 *The graphs K_5 and $K_{3,3}$ are not planar.*

Proof For K_5 we have $|X| = 5$ vertices and $|E| = 10$ edges, so

$$|E| = 10 > 9 = 3|X| - 6.$$

The graph $K_{3,3}$ is bipartite and therefore contains no triangles. There are $|X| = 6$ vertices and $|E| = 9$ edges and

$$|E| = 9 > 8 = 2|X| - 4.$$

So in both cases, the necessary conditions for planarity (Theorem 4.1.3) are not fulfilled. $\qquad\square$

Actually, the graphs K_5 and $K_{3,3}$ play a very special role. Namely, the following planarity criterion holds [73]:

Theorem 4.2.2 (Kuratowski) *A graph is planar if and only if none of its subgraphs is a subdivision of K_5 or $K_{3,3}$.*

Before we get to the proof, we would like to reformulate this important theorem in the language of minors [111]: The new version will be important in Sect. 4.5. As a reminder, all minors of a graph are obtained by successively removing vertices and edges and by applying edge contractions (Definition 1.1.25).

Theorem 4.2.3 (Wagner) *A graph is planar if and only if K_5 and $K_{3,3}$ are not among its minors.*

Proof If G is planar, then all minors of G are planar. Since K_5 and $K_{3,3}$ are non-planar, they cannot be minors of G.

Assume that G is not planar. According to Kuratowski's theorem, G contains a subgraph G', which is a subdivision of K_5 or $K_{3,3}$. In particular, G' is also a minor of G. Now in G' all "artificial" edges e (those with $\deg \iota e = 2$ or $\deg \tau e = 2$) can be contracted one after the other, then K_5 or $K_{3,3}$ is obtained as a minor of G' and thus also of G. $\qquad\square$

In the rest of the section, we will deal with the proof[1] of Kuratowski's theorem (Theorem 4.2.2). Our proof will be by contradiction. By a **Kuratowski subgraph** of G we mean a subgraph of G that is a subdivision of K_5 or $K_{3,3}$.

Lemma 4.2.4 *Let G be a graph such that*

(a) G is non-planar,
(b) G does not contain a Kuratowski subgraph,
(c) the number of edges in G is minimal (among all graphs with these properties),

then G is 3-connected.

For the proof, we need the following observation, which will also play a role later.

---

[1] Note for the lecturers: Kuratowski's theorem can also be accepted without proof, as the exact proof constructions will not be used in the later text.

Lemma 4.2.5 *Given a planar drawing of a graph H and $e_1, \ldots, e_n$ are the boundary edges of a face f. Then there is a planar drawing of H such that $e_1, \ldots, e_n$ are the boundary edges of the unbounded face.*

Proof We identify $\mathbb{R}^2$ with $\mathbb{C}$. Let z_0 be any point within the face f, then the mapping $z \mapsto 1/(z - z_0)$ transforms the existing drawing into another planar drawing, thereby mapping the face f onto the unbounded face (Fig. 4.4). This proves the statement. $\square$

Proof of Lemma 4.2.4 We prove an auxiliary statement

(H) Let H be a minimal non-planar graph (i.e., H is non-planar, but every proper subgraph of H is planar). Then H is 2-connected.

If H were not connected, each component of H would have to be planar (as a proper subgraph of H), but then H would also be planar. Therefore, H must be connected. Assume that there is a vertex x such that $H \backslash x$ is not connected. Consider the components $G_1, \ldots, G_n$ of $H \backslash x$. We construct new graphs H_i by supplementing G_i with the vertex x and all edges of H between G_i and x (Fig. 4.5).

Then all H_i are proper subgraphs of H and thus planar. According to Lemma 4.2.5, for each H_i we can find a planar drawing such that x is on the boundary of the unbounded face. This allows us to glue the drawings of H_i into a planar drawing of H (the drawings of the H_i can be suitably scaled and rotated), so H is also planar. Therefore, we must remove at least two vertices in H to make it disconnected, so H is 2-connected. This shows (H).

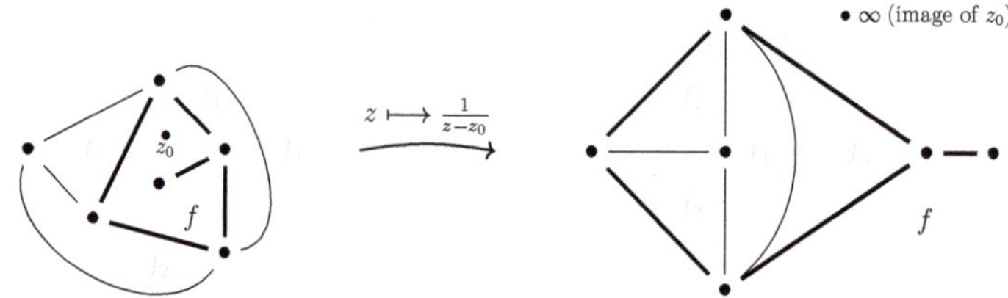

Fig. 4.4 Construction of a new planar drawing

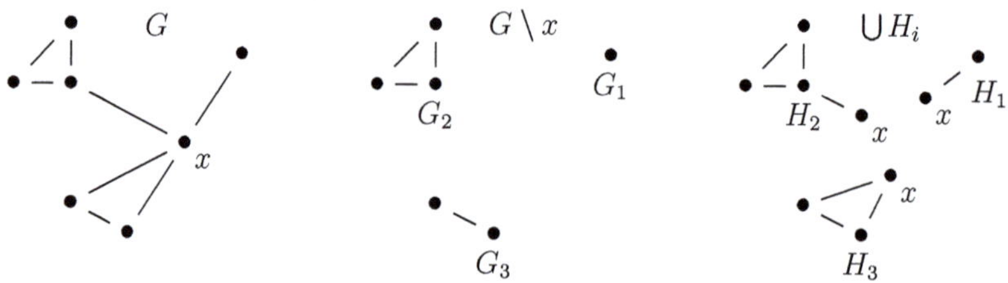

Fig. 4.5 Construction of the subgraphs H_i

Now we prove the statement of the lemma. Let $G = (X, E)$ be a graph without Kuratowski subgraphs and such that the number of edges is minimal. Let $e \in E$, then $G \setminus e$ also does not contain any Kuratowski subgraphs and due to the minimality assumption (c), $G \setminus e$ must be planar. Thus, we have shown that G is a minimal non-planar graph, and by (H) it is at least 2-connected.

Assume that G is not 3-connected. Then there exist $x, y \in X$, with $x \neq y$, such that $G \setminus \{x, y\}$ is not connected. Let $G_1, \ldots, G_n$ be the connected components of $G \setminus \{x, y\}$ and construct new graphs H_i by supplementing G_i with the vertices x and y and the edges of G between G_i and x and y. Then consider the graphs

$$\widetilde{H}_i := H_i \cup \text{edge } \{x, y\}$$

(it should be noted that we do not yet know whether G contains the edge $\{x, y\}$ or not). If all $\widetilde{H}_i$ are planar, they can be drawn in the plane without crossings according to Lemma 4.2.5 such that the edge $\{x, y\}$ always lies on the outer boundary. Then the drawings can be adjusted and glued together along $\{x, y\}$ to create a planar drawing of $G' := G \cup \text{edge } \{x, y\}$: First draw $\widetilde{H}_1$, then draw $\widetilde{H}_2$ within a face of $\widetilde{H}_1$ and so on (Fig. 4.6).

Therefore, G' is planar and thus $G' \supset G$ is also a planar graph: contradiction. Hence, there exists an $i \in \{1, \ldots, n\}$ such that $\widetilde{H}_i$ is not planar.

For each j, there is at least one edge between x and G_j and at least one edge between y and G_j (otherwise, G would already decompose into several connected components after the removal of only x or only y and would therefore not be 2-connected). Since H_j with different j have no common edges, each H_j contains at most $|E| - 2$ edges. Then, the graph $\widetilde{H}_i$ (for the i chosen above) has at most $|E| - 1$ edges (thus fewer than G) and due to the minimality assumption (c), $\widetilde{H}_i$ must contain a Kuratowski subgraph F.

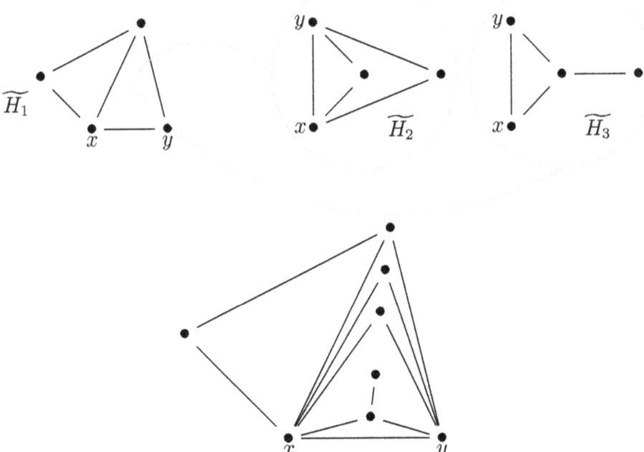

Fig. 4.6 Construction of the subgraphs $\widetilde{H}_i$

If the original graph G contains the edge $\{x, y\}$, then $\widetilde{H}_i$ is a subgraph of G and thus G contains the Kuratowski subgraph F: contradiction!

Now consider the case $\{x, y\} \notin E$. As noted above, x and y have neighbors in each G_j. Since G_j is connected, x and y can be connected by a path P_j in G that is entirely within H_j. Then choose any $j \neq i$ and replace in F the edge $\{x, y\}$ with the path P_j: Thus, we have found another Kuratowski subgraph in G: contradiction!

Therefore, G must necessarily be 3-connected. $\qquad\square$

Actually, we are going to show that graphs described in Lemma 4.2.4 do not exist (from which a contradiction will arise). However, this requires significantly more work.

Lemma 4.2.6 *If $G = (X, E)$ is a 3-connected graph with at least 5 vertices, then for each edge $e \in E$ the graph $G \cdot e$ is at least 2-connected.*

Proof Denote by x_e the new vertex that arises when contracting an edge $e = \{x, y\} \in E$. First note that $G \cdot e$ always remains connected, i.e., $G \cdot e$ is at least 1-connected. If v is a vertex in $G \cdot e$, then

$$(G \cdot e) \setminus v = \begin{cases} (G \setminus v) \cdot e, & v \neq x_e, \\ G \setminus \{x, y\}, & v = x_e, \end{cases}$$

and by assumption, $G \setminus v$ and $G \setminus \{x, y\}$ are connected, so $(G \cdot e) \setminus v$ is also connected. Thus, we have shown that $G \cdot e$ is at least 2-connected. $\qquad\square$

The last statement can be strengthened:

Lemma 4.2.7 *If $G = (X, E)$ is a 3-connected graph with at least 5 vertices, then there exists an edge $e \in E$ such that $G \cdot e$ is again 3-connected.*

Proof According to Lemma 4.2.6, $G \cdot e$ is at least 2-connected. Let x_e be the new vertex resulting from the contraction of an edge $e = \{x, y\} \in E$. We will conduct a proof by contradiction.

Assume that for all $e = \{x, y\} \in E$, the graph $G \cdot e$ is not 3-connected, then there are always two vertices u, v in $G \cdot e$ such that $(G \cdot e) \setminus \{u, v\}$ is not connected. If $x_e \notin \{u, v\}$, then u, v are vertices in G and $G \setminus \{u, v\}$ is not connected: contradiction to the assumption that G is 3-connected. Therefore, either $x_e = u$ or $x_e = v$ must hold. Denote by z the other vertex, i.e. $\{z\} = \{u, v\} \setminus x_e$, then $(G \cdot e) \setminus \{u, v\} = G \setminus \{x, y, z\}$.

Thus, we have shown the following:

$$\text{For each edge } e = \{x, y\} \text{ in } G, \text{ there is a vertex } z \text{ in } G, \text{ with} \\ z \notin \{x, y\}, \text{ such that } G \setminus \{x, y, z\} \text{ is not connected.} \tag{4.3}$$

Now we choose $e = \{x, y\}$ and the corresponding vertex z such that $G \setminus \{x, y, z\}$ contains a connected component H with *minimal* number of vertices. Each of the vertices x, y, z has at least one neighbor in each connected component of $G \setminus \{x, y, z\}$ (otherwise, G would already decompose into several connected components after the removal of fewer than 3 vertices: impossible, since G is 3-connected). Let u be a vertex in H such that $z \sim u$ in G (in particular $u \notin \{x, y\}$). According to (4.3), there is another vertex v in G such that $G \setminus \{z, u, v\}$ is not connected. Two cases are possible:

- One of the connected components of $G \setminus \{z, u, v\}$ contains the edge $\{x, y\}$: This corresponds to the case $v \notin \{x, y\}$,
- One of the vertices x, y is v. In this case, the other vertex is in a connected component of $G \setminus \{z, u, v\}$. Therefore, there is another connected component that contains neither x nor y.

In both cases, there is a connected component H' of $G \setminus \{z, u, v\}$ that contains neither x nor y. By construction, z is also not contained in H', so H' is a connected subgraph of $G \setminus \{x, y, z\}$, see Fig. 4.7.

The graph H' contains at least one neighbor u' of u (otherwise G would already fall apart into several parts after the removal of z and v: impossible due to the 3-connectivity). Thus, we have shown that H and H' are connected by an edge. Since H is a maximal connected subgraph of $G \setminus \{x, y, z\}$, it follows that $H' \subset H$. But $u \in H \setminus H'$ by construction, i.e., H' has strictly fewer vertices than H, which contradicts the choice of H. $\qquad\square$

Lemma 4.2.8 *Let $G = (X, E)$ be a graph and let $e \in E$ such that $G \cdot e$ contains a Kuratowski subgraph. Then G also contains a Kuratowski subgraph.*

Proof Let $e = \{x, y\}$ and x_e be the new vertex that arises when contracting e. Let H be a Kuratowski subgraph of $G \cdot e$. If x_e is not in H, then H is a subgraph of G and the statement is true. From now on, we assume that H contains the vertex x_e.

A vertex in a subdivision of K_5 or $K_{3,3}$ can only have degree 2, 3 or 4, so x_e has either 2, 3 or 4 neighbors in H. We consider the following situations:

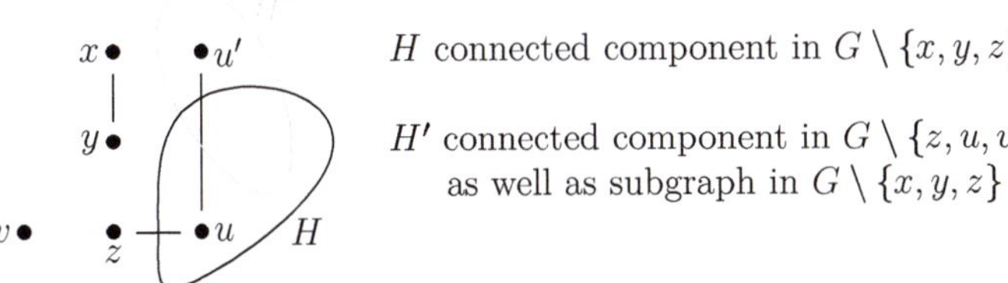

H connected component in $G \setminus \{x, y, z\}$

H' connected component in $G \setminus \{z, u, v\}$
as well as subgraph in $G \setminus \{x, y, z\}$

Fig. 4.7 The subgraphs H and H'

(a) All neighbors of x_e in H are neighbors of x in G. Then one obtains a Kuratowski subgraph of G by replacing x_e with x (Fig. 4.8).

(b) All neighbors of x_e in H are neighbors of y in G. Then a Kuratowski subgraph of G is obtained by replacing x_e with y.

(c) There is a neighbor u of x_e in H, who is a neighbor of x in G, all other neighbors of x_e in H are neighbors of y in G. Replace the edge $\{x_e, u\}$ from H with the path (y, x, u) to obtain a Kuratowski subgraph in G (Fig. 4.9).

(d) There is a neighbor u of x_e in H, who is a neighbor of y in G, all other neighbors of x_e in H are neighbors of x in G. Replace the edge $\{x_e, u\}$ from H with the path (x, y, u) to obtain a Kuratowski subgraph in G.

(e) There are two neighbors u_1, u_2 of x_e in H, who are neighbors of x in G, and there are two other neighbors u_3, u_4 of x_e in H, who are neighbors of y in G. Since x_e in H has a total of 4 neighbors, H is a subdivision of K_5 (in a subdivision of $K_{3,3}$ the maximum degree is 3). Then H contains four additional vertices v_1, v_2, v_3, v_4, which have degree 4 in H, while all remaining vertices in H have degree 2. We can assume w.l.o.g. that u_j lies on the path from x_e to v_j (Fig. 4.10). Since all u_j are also directly connected to each other (without having to go through x_e), we obtain in G a subgraph that contains x and y as well as the vertices v_j and is a subdivision of $K_{3,3}$.

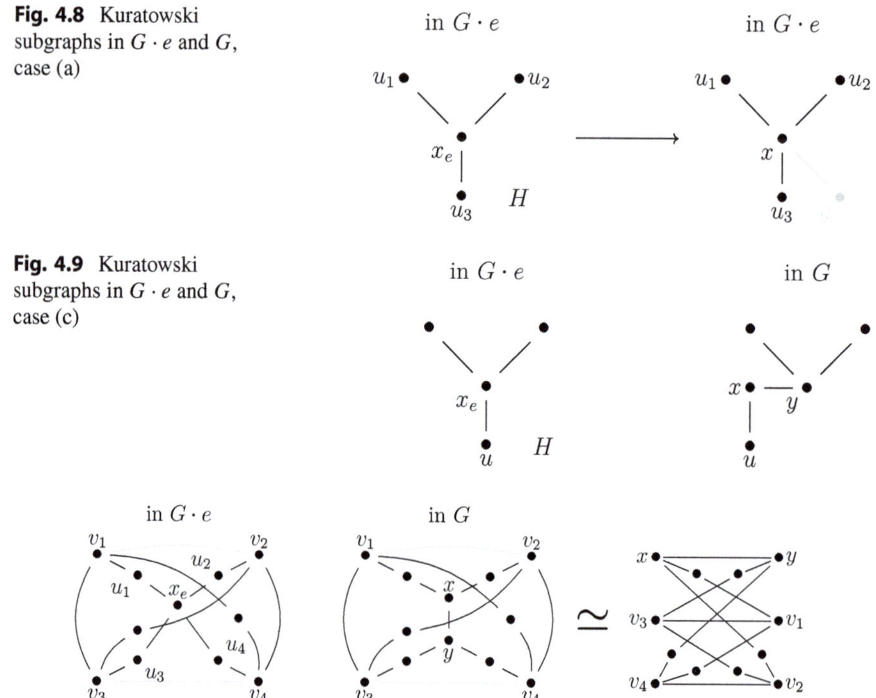

Fig. 4.8 Kuratowski subgraphs in $G \cdot e$ and G, case (a)

Fig. 4.9 Kuratowski subgraphs in $G \cdot e$ and G, case (c)

Fig. 4.10 Kuratowski subgraphs in $G \cdot e$ and G, case (e)

We will now check whether we have considered all possible situations.

- x_e has exactly two neighbors u_1 and u_2 in H.
  - If both u_1 and u_2 are neighbors of x in G or neighbors of y in G, we are in case (a) or (b), respectively.
  - Otherwise, one of the vertices u_1, u_2 is a neighbor of x in H and the other vertex is a neighbor of y in G, so we are in case (c).
- x_e has exactly three neighbors u_1, u_2, u_3 in H.
  - If all u_j are neighbors of x in G or neighbors of y in G, we are in case (a) or (b), respectively.
  - If only one of the u_j is a neighbor of x in G, the other two vertices must be neighbors of y in G and we are in case (c).
  - If exactly two of the u_j are neighbors of x in G, the third vertex must be a neighbor of y in G, so we are in case (d).
- x_e has exactly four neighbors u_1, u_2, u_3, u_4 in H.
  - If all u_j are neighbors of x in G or neighbors of y in G, we are in case (a) or (b), respectively.
  - If only one of the u_j is a neighbor of x in G, the other three vertices must be neighbors of y in G and we are in case (c).
  - If exactly two of the u_j are neighbors of x in G, the remaining two vertices must be neighbors of y in G, so we are in case (e).
  - If exactly three of the u_j are neighbors of x in G, the fourth vertex must be a neighbor of y in G, so we are in case (d).

Thus, we have covered all possible cases. $\qquad\qquad\qquad\qquad\qquad\qquad\square$

Lemma 4.2.9 *Every 3-connected graph that does not contain a Kuratowski subgraph is planar.*

Proof We use induction on the number of vertices. The only 3-connected graph with 4 vertices is K_4, and it is planar. Let now $n \geq 5$ such that all 3-connected graphs with at most $n - 1$ vertices and without Kuratowski subgraphs are planar.

Let G be a 3-connected graph with n vertices and without Kuratowski subgraphs. According to Lemma 4.2.7, there is an edge $e = \{a, b\}$ in G such that $G \cdot e$ is again 3-connected. According to Lemma 4.2.8, $G \cdot e$ also does not contain any Kuratowski subgraphs. Since $G \cdot e$ has only $n-1$ vertices, it is planar by the induction hypothesis.

Let x_e be the vertex that was created by the contraction of e. Consider a planar drawing of $G \cdot e$. When removing x_e with all incident edges, we obtain a planar drawing of

$$G' := (G \cdot e) \setminus x_e \equiv G \setminus \{a, b\}.$$

Let f be the face in this drawing that contains x_e. Since $G \cdot e$ is 3-connected, G' is at least 2-connected, so f is bordered by a cycle $C = (u_0, u_1, \ldots, u_m = u_0)$ (Lemma 4.1.1), so $m \geq 3$. This cycle is also a cycle in the original graph G, as it does not contain x_e. By construction, the cycle C contains all neighbors of a and b. Now, it is only a matter of drawing the vertices a and b with the corresponding edges inside the face f without crossings, see Fig. 4.11.

We assume w.l.o.g. that

$$\deg a \ \leq \ \deg b. \tag{4.4}$$

(otherwise swap the roles of a and b).

Let $a_1, \ldots, a_k$ be the neighbors of a in cyclic order (we set $a_0 := a_k$), then $\{a_1, \ldots, a_k\} \subset C$ and we denote by P_j the path from a_{j-1} to a_j along C, see Fig. 4.12. Every vertex of C lies in at least one of the P_j.

We draw a arbitrarily within f and connect it through suitable curves γ_j without crossings to the a_j. Now b and the corresponding edges need to be drawn. We note that the case $\deg a = 1$ is impossible (since in this case only a and b would be connected and the graph G would not be connected). Now we distinguish between several cases:

(i) $\deg a = 2$. In this case, a is only connected to b and to another vertex of C, and we can place b arbitrarily within the face and connect it without crossings to all vertices on C, so G is planar, see Fig. 4.13.

From now on, we assume that $\deg a \geq 3$.

(ii) a and b have at least three common neighbors.

Let u, v, w be three different common neighbors of a and b. If one considers all edges between a, b, u, v, w as well as the paths between u, v, w over C, one obtains a subdivision of K_5 (Fig. 4.14): contradiction.

(iii) All neighbors of b (except a) are also neighbors of a.

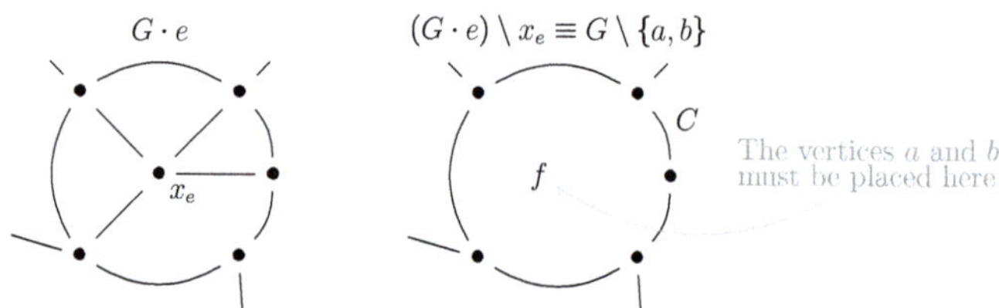

Fig. 4.11 The cycle C, the face f and the vertices a, b, x_e

Fig. 4.12 The paths P_j

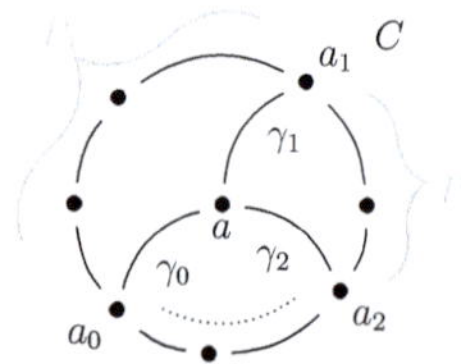

Fig. 4.13 Placement of a and b in case (i)

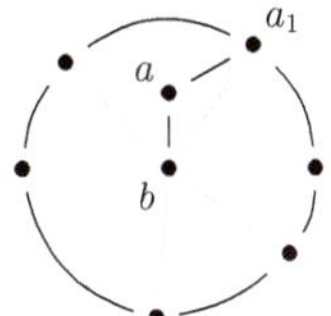

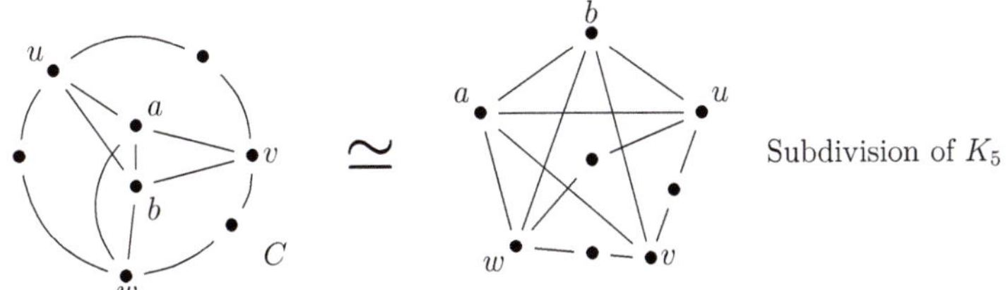

Fig. 4.14 Kuratowski subgraphs in case (ii)

Fig. 4.15 Placement of a and b in case (iv)

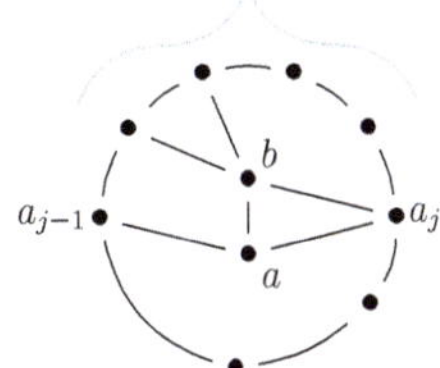

From (4.4) it follows that $\deg a = \deg b$. From the assumption $\deg a \geq 3$ it follows that a and b have at least three common neighbors: This is already covered by (ii).

(iv) For some j all neighbors of b (except a) are in P_j.

Draw b arbitrarily within the triangular face f', which is bounded by P_j and the curves γ_{j-1} and γ_j. Then one can connect b with all its neighbors without crossings by curves within f' (Fig. 4.15). This results in a planar drawing of G, so G is planar.

(v) b has exactly one neighbor $u \neq a$, who is not a neighbor of a. Then $\deg b \leq \deg a + 1$, so overall

$$3 \leq \deg a \leq \deg b \leq \deg a + 1. \tag{4.5}$$

Then there is a uniquely determined j with $u \in P_j$. We can assume w.l.o.g. $j = 1$.

(v.1) $\deg a = 3$ (then there are only two paths P_1 and P_2). Then all further neighbors of b are among a_1 and a_2: Both are in P_1, so we are in case (iv).

(v.2) $\deg b \geq 5$. All neighbors of b except a and u are neighbors of a, then a and b have at least 3 common neighbors: Case (ii).

(v.3) $\deg a \geq 5$. From (4.5) it follows $\deg b \geq 5$ and we are in case (v.2).

(v.4) $\deg a = \deg b = 4$.

Since we already have two neighbors a and u of b, two more neighbors are among the vertices a_0, a_1, a_2.

(v.4.1) $b \sim a_0$ and $b \sim a_1$. Then all neighbors of b (except a) are in P_1 and we are in case (iv), see Fig. 4.16.

(v.4.2) b is adjacent to only one of a_0, a_1. W.l.o.g. let $b \not\sim a_0$, then $b \sim a_1$, $b \sim a_2$. If we contract the edge $e' := \{a_0, u\}$ to a new vertex z, z, a_1, a_2, a, b are connected by paths and thus form a subgraph in $G \cdot e'$, which is a subdivision of K_5 (Fig. 4.17). According to Lemma 4.2.8, G contains a Kuratowski subgraph: contradiction.

(vi) b has at least two neighbors u, v, which are different from a and are not neighbors of a.

(vi.1) One has $u, v \in P_j$ for some j.

- If all other neighbors of b are in P_j, then we are in case (iv).6
- If b has another neighbor $w \notin \{a, u, v\}$ with $w \notin P_j$, then we have a subgraph of G constructed on a_{j-1}, u, a_j, w, a, b which is a subdivision of $K_{3,3}$ (Fig. 4.18): contradiction.

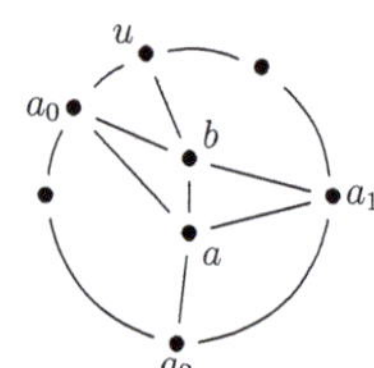

Fig. 4.16 Placement of a and b in case (v.4.2)

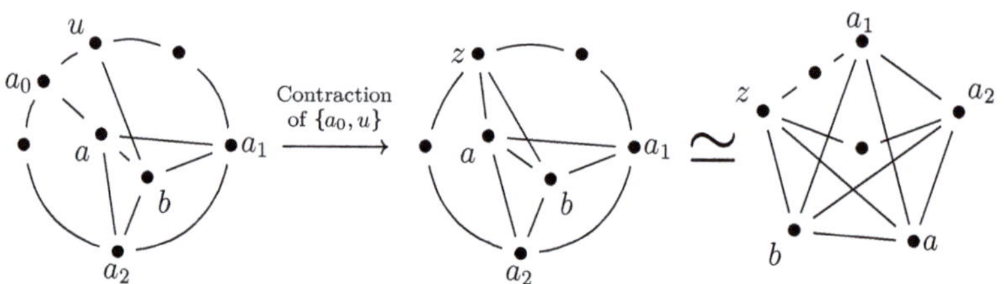

Fig. 4.17 Placement of a and b and Kuratowski subgraphs in case (v.4.2)

Fig. 4.18 Kuratowski subgraphs in case (vi.1)

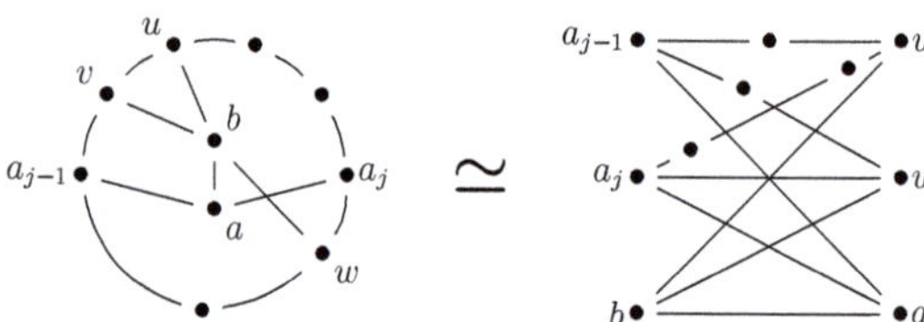

Fig. 4.19 Kuratowski
subgraphs in case (vi.2)

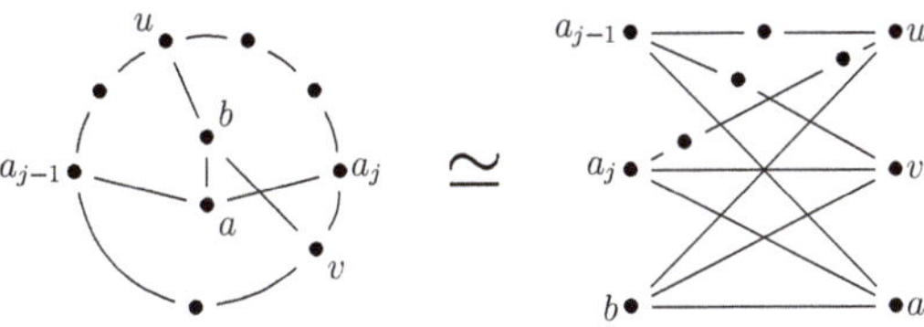

(vi.2) There is no j with $u, v \in P_j$.

There is a uniquely determined j with $u \in P_j$, then a_{j-1}, u, a_j and v are four different vertices that stand on C in exactly this order and together with a and b they form a subgraph in G that is a subdivision of $K_{3,3}$ (Fig. 4.19): contradiction.

Thus, we have shown that G is planar in all possible situations.   □

Proof of Theorem 4.2.2 (Kuratowski's Theorem) Let G be a planar graph. Then all subgraphs of G are planar. Since subdivisions of K_5 and $K_{3,3}$ are not planar, they cannot occur as subgraphs of G.

For the other direction, we assume that non-planar graphs without Kuratowski subgraphs exist. Then there is such a graph G in which the number of edges is minimal. According to Lemma 4.2.4, G is 3-connected, and Lemma 4.2.9 shows that G is planar: contradiction.   □

4.3 Maximal Planar Graphs

As a preparation for the next topic, we consider a special class of planar graphs. A planar graph G is called **maximal planar**, if no additional edge can be added without losing planarity. Every maximal planar graph is automatically connected (otherwise, two connected components could be connected by an additional edge: the planarity would not be affected). But much stronger statements can be proven.

Lemma 4.3.1 *Every maximal planar graph with at least 3 vertices is 2-connected.*

Proof Let $G = (X, E)$ be a maximal planar graph with $|X| \geq 3$. As already noted, G is connected. Assume that G is not 2-connected, then there is an $x \in X$, such that $G' := G \setminus x$ is not connected. Let $G'_1, \ldots, G'_k$ be the connected components of $G \setminus x$ (with $k \geq 2$). By Lemma 4.2.5, there is a crossing-free drawing of G such that x is a boundary vertex for the unbounded face of G. Then x lies within the unbounded face f' of $G \setminus x$ for the same drawing, and all neighbors of x lie on the boundary of f'. Then there are two boundary vertices $y_j \in G'_j$ and $y_k \in G'_k$ of f' with $y_j \nsim y_k$ but $y_j \sim x \sim y_k$. Then the edge $\{y_j, y_k\}$ can be added without losing planarity: contradiction to the maximal planarity of G.   □

Lemma 4.3.2 *Let $G = (X, E)$ be a planar graph with at least 3 vertices, then the following three conditions are equivalent:*

(a) G is maximal planar,
(b) G is a triangle graph, i.e., every face in every crossing-free drawing is bordered by a cycle of length 3,
(c) $|E| = 3|X| - 6$.

Proof (a) $\Rightarrow$(b)   Let G be maximal planar and consider a crossing-free drawing of G. Since G is 2-connected, every face is bordered by a cycle (Lemma 4.1.1). Suppose that G is not a triangle graph, then there is a face f bordered by a cycle $(x_0, x_1, \ldots, x_m = x_0)$ of length $m \geq 4$. If there are two j, k with $x_j \neq x_k$ and $x_j \not\sim x_k$, one can draw a smooth curve from x_j to x_k within f. Thus, the graph remains planar when the additional edge $\{x_j, x_k\}$ is added (Fig. 4.20): contradiction to the maximal planarity.

Therefore, all x_j are connected by edges. If $m \geq 5$, K_5 is a subgraph of G and according to Kuratowski's theorem, G is not planar: contradiction. Therefore, $m = 4$. Note that x_1 lies on the closed curve $\{x_0, x_1\} \cup \{x_1, x_2\} \cup \{x_2, x_0\}$ and the edge from x_1 to x_3 must intersect this curve (since this edge does not run over f by assumption, see Fig. 4.21): contradiction to planarity. Therefore, G is a triangular graph.

(b) $\Rightarrow$(c)   Consider a crossing-free drawing of G and let F be the set of faces. Since each edge lies on the boundary of exactly two faces and each face is bordered by 3 edges, we have $3|F| = 2|E|$ and the statement follows from Euler's formula $|X| + |F| - |E| = 2$.

(c) $\Rightarrow$(a)   Since $3|X| - 6$ is the maximum number of edges in a planar graph (Theorem 4.1.3), the planarity is destroyed by adding any additional edge, so G is maximal planar.

$\square$

Theorem 4.3.3 *Any maximal planar graph with at least 4 vertices is 3-connected.*

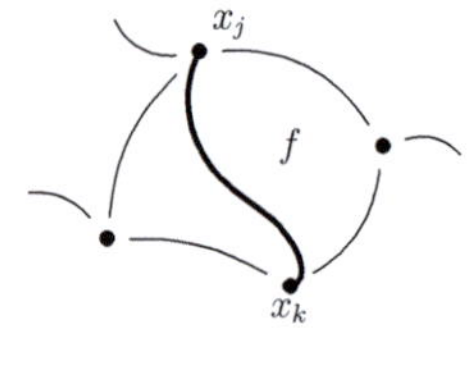

Fig. 4.20 Preservation of planarity when adding the edge $\{x_j, x_k\}$

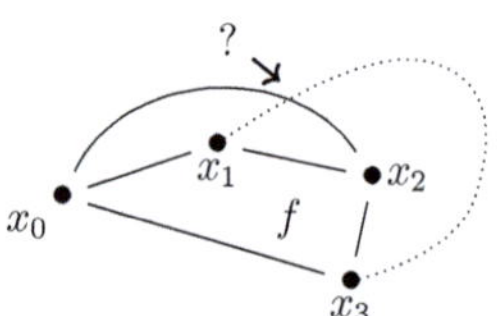

Fig. 4.21 Non-existence of the edge $\{x_1, x_3\}$

Fig. 4.22 The cycle C_u

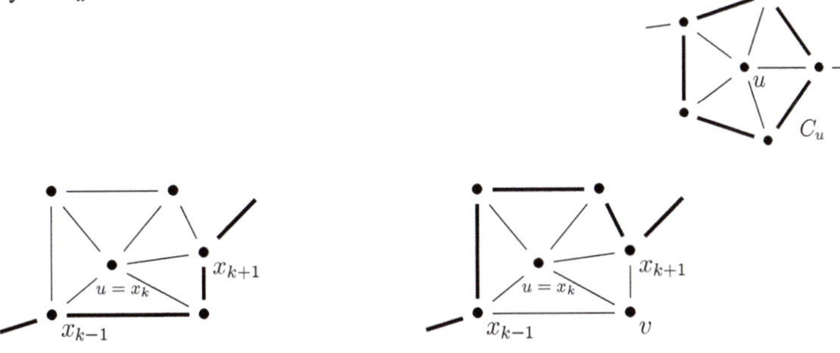

Fig. 4.23 Possible paths over x_{k-1} and x_{k+1}

Proof Let $G = (X, E)$ be a maximal planar graph with $|X| \geq 4$. Let $u, v \in X$ with $u \neq v$. We need to show that $G' := G \setminus \{u, v\}$ is connected. For this, we consider an arbitrary crossing-free drawing of G. Let $u_0, \ldots, u_{p-1}$ be the neighbors of u in cyclic order. Since G is a triangular graph, they form a cycle $C_u := (u_0, u_1, \ldots, u_p = u_0)$, see Fig. 4.22. Similarly, there is a cycle $C_v = (v_0, v_1, \ldots, v_q = v_0)$ that runs over all neighbors of v. We note that there are exactly two paths between two vertices in a cycle.

Let $x, y \in X \setminus \{u, v\}$ with $x \neq y$. Since G is connected, there is a path $(x = x_0, x_1, \ldots, x_n = y)$. We now make a case distinction.

- The vertices u and v are not included in the path. Then x, y are also connected in G' by a path.
- The path contains exactly one of the vertices u, v, e.g. u. Let k with $x_k = u$, then x_{k-1} and x_{k+1} lie on the cycle C_x and one can replace the subpath $(x_{k-1}, x_k = u, x_{k+1})$ by the path from x_{k-1} to x_{k+1} over C_u (if v also lies on C_x, one can choose the path that does not run over v), see Fig. 4.23. This results in a path from x to y that contains neither u nor v, so x, y are also connected in G' by a path.
- The path contains both u and v. W.l.o.g. we assume that one first runs over u and then over v: $x_k = u$ and $x_m = v$ with $k < m$. If $u \sim v$, one replaces the original subpath from $(x_{k-1}, \ldots, x_{m+1})$ by a path over C_u and C_v.
  If $u \not\sim v$, then one replaces the subpath $(x_{k-1}, x_k = u, x_{k+1})$ by the path from x_{k-1} to x_{k+1} over the cycle C_u and the subpath $(x_{m-1}, x_k = v, x_{m+1})$ by the path from x_{m-1} to x_{m+1} over the cycle C_v.
  So in this case too, there is a path from x to y in G'.

With this, we have shown that G' is connected.   □

Exercise 4.3.1 A graph is called *outerplanar*, if it is planar <u>and</u> it can be drawn without crossings such that all vertices lie on the boundary of the unbounded face.

An outerplanar graph is called *maximal outerplanar* if no further edge can be added without losing the outerplanarity.

Let G be a graph and

$$\widetilde{G} := G \cup \text{ a new vertex connected to all other vertices.}$$

Show:

1. G is outerplanar if and only if $\widetilde{G}$ is planar.
2. G is outerplanar if and only if G does not contain a subdivision of K_4 or $K_{3,2}$.
3. If G is outerplanar, then $\chi(G) \le 3$.
4. Every maximal outerplanar graph is connected.
5. G is maximal outerplanar if and only if $\widetilde{G}$ is maximal planar.

4.4   Colin de Verdière Invariant

As a reminder: Weighted Laplace matrices were defined in Sect. 3.3 (see Definition 3.3.1). From now on, we will only work with real vector spaces.

Definition 4.4.1 Let $G = (X, E)$ be a graph. A **Colin de Verdière matrix** (short: CdV Matrix) for G is a weighted Laplace matrix $M = (M_{x,y})_{x,y \in X}$ for G with the following additional properties:

(M1) M has exactly one negative eigenvalue,
(M2) The only real symmetric matrix $R = (R_{x,y})_{x,y \in X}$ with

$$R_{x,y} = 0 \text{ for all } x, y \text{ with } x = y \text{ or } x \sim y \tag{4.6}$$

<u>and</u> $MR = 0$ is the zero matrix.

The set of all CdV matrices for G is denoted by $\mathrm{CdV}(G)$.

Proposition 4.4.2 *For every graph G there is at least one CdV matrix, so* $\mathrm{CdV}(G)$ *is not empty.*

Proof Let $G_1, \ldots, G_n$ be the connected components of G and $L_1, \ldots, L_n$ their Laplace matrices. If G_1 contains at least two vertices, then $\lambda_2(L_1) > 0$, so we choose an $a \in \bigl(0, \lambda_2(L_1)\bigr)$. If G_1 consists of only one vertex, then $L_1 = (0)$, so in this case we choose any $a > 0$. In both cases, $(-a)$ is the only negative eigenvalue of $L_1 - aI$. Now consider the block matrix

$$M := (L_1 - aI) \oplus (L_2 + I) \oplus \cdots \oplus (L_n + I).$$

This is clearly a weighted Laplace matrix for G. For all $k \in \{2, \ldots, n\}$, all eigenvalues of $L_k + I$ are positive, so M has the only negative eigenvalue $(-a)$, i.e., (M1) is fulfilled. Furthermore, $L_k + I$ as well as $L_1 - aI$ are invertible, so M is also invertible, and from $MR = 0$ it follows $R = 0$: Thus, property (M2) is also fulfilled. $\qquad\qquad\square$

Definition 4.4.3 Let G be a graph. The **Colin de Verdière invariant $\mu(G)$** (short: CdV Invariant) of G is

$$\mu(G) := \max\left\{\dim \ker M : M \in \mathrm{CdV}(G)\right\}.$$

Remark 4.4.4 The invariant $\mu(G)$ was originally introduced in [30]. Our presentation mainly follows the later overview article [108].

Remark 4.4.5 Since for $G = (X, E)$ each CdV matrix of G has exactly $|X|$ eigenvalues and one of them is negative, the trivial estimate

$$0 \le \mu(G) \le |X| - 1.$$

holds.

In the practical calculation of $\mu(G)$, condition (M2) is the hardest to check. The following remark simplifies some proofs:[2]

Lemma 4.4.6 *Let $G = (X, E)$ be a graph and M a weighted Laplace matrix for G which fulfills condition* (M1) *in Definition 4.4.1* <u>and</u> $\dim \ker M \le 1$. *Then M also fulfills* (M2) *and thus $M \in \mathrm{CdV}(G)$.*

Proof If $\dim \ker M = 0$, then M is invertible, so from $MR = 0$ it immediately follows $R = 0$ and (M2) holds.

Now let $\dim \ker M = 1$. Assume that there exists a real symmetric matrix R with property (4.6) such that $MR = 0$ but $R \ne 0$. From $MR = 0$ it follows that $\mathrm{ran}\, R \subset \ker M$ and $\dim \mathrm{ran}\, R \le 1$. For $\dim \mathrm{ran}\, R = 0$, $R = 0$ would hold, so consider the case $\dim \mathrm{ran}\, R = 1$. Then $\mathrm{ran}\, R = \mathbb{R}v$ for a $v \in \mathbb{R}^X$ with $v \ne 0$ and

$$R : \mathbb{R}^X \ni f \longmapsto a(f)v \in \mathbb{R}^X$$

for a linear mapping $a : \mathbb{R}^X \to \mathbb{R}$. The mapping a has the form $a : f \mapsto \langle u, f \rangle$ with $u \ne 0$ (for $u = 0$ we would have $R = 0$), so

$$R : f \mapsto \langle u, f \rangle v, \qquad R^* : f \mapsto \langle v, f \rangle u.$$

---

[2] Note for the lecturers: Lemma 4.4.6 is not used in the main text, but is important for the exercises.

From $R = R^*$ it follows that $u = \lambda v$ for a $\lambda \in \mathbb{R} \setminus \{0\}$, so $R : f \mapsto \lambda \langle v, f \rangle v$, and for each $x \in X$ it holds that

$$(Rf)(x) = \lambda \Big(\sum_{y \in X} v(y) f(y) \Big) v(x) = \sum_{y \in X} \lambda v(x) v(y) \, f(y),$$

i.e. $R = (R_{x,y})$ with $R_{x,y} = \lambda v(x) v(y)$. From (4.6) it follows that $R_{x,x} = 0$ for all $x \in X$ and thus $v = 0$ and $R = 0$: contradiction. Therefore, (M2) is satisfied for M. $\qquad\square$

With the above preparations, we can calculate $\mu(G)$ for some important graphs G.

Example 4.4.7

(a) Consider $G = K_n$. The matrix $M := -J_n$ is a weighted Laplace matrix for K_n. It has the simple eigenvalue $(-n) < 0$ as well as the $(n-1)$-fold eigenvalue 0 (or no further eigenvalues, if $n = 1$), so it satisfies (M1). The only matrix R that satisfies the conditions (4.6) is the zero matrix, therefore, M also satisfies (M2), so $M \in \mathrm{CdV}(K_n)$ and thus $\mu(K_n) \geq \dim \ker M = n - 1$. Since $\mu(G) \leq n - 1$ for all graphs with n vertices (Remark 4.4.5), we obtain

$$\mu(K_n) = n - 1 \text{ for all } n \in \mathbb{N}.$$

(b) Let $G = K_{m,n}$. The block matrix

$$M = \begin{pmatrix} 0 & -J_{m,n} \\ -J_{n,m} & 0 \end{pmatrix}$$

is obviously a weighted Laplace matrix for $K_{m,n}$. We have $M = -A$, where A is the adjacency matrix of $K_{m,n}$, for which we have calculated the eigenvalues in Example 1.6.7. Thus, M has the simple eigenvalues $\pm\sqrt{mn}$ as well as the $(m+n-2)$-fold eigenvalue 0. With this, we see that M satisfies condition (M1). With (M2) it looks more complicated this time. The real symmetric matrices R that satisfy (4.6) are the block matrices

$$R = \begin{pmatrix} Y & 0 \\ 0 & Z \end{pmatrix}$$

with real symmetric $m \times m$ matrices Y and real symmetric $n \times n$ matrices Z, both with zeros on the diagonal. The condition $MR = 0$ has the form

$$\begin{pmatrix} 0 & J_{m,n}Z \\ J_{n,m}Y & 0 \end{pmatrix} = 0$$

and is equivalent to

The sum of the coefficients in each row and each column of Y and Z is 0.

$$(4.7)$$

In general, this does *not* imply that $R = 0$. For example, for $m = 4$ one can take

$$R = \begin{pmatrix} Y & 0 \\ 0 & 0 \end{pmatrix} \quad \text{with } Y = \begin{pmatrix} 0 & -1 & 0 & 1 \\ -1 & 0 & 1 & 0 \\ 0 & 1 & 0 & -1 \\ 1 & 0 & -1 & 0 \end{pmatrix}$$

So in general, M does <u>not</u> satisfy condition (M2) and is therefore not a CdV matrix for $K_{m,n}$.

For $m, n \leq 3$, however, one can show that (4.7) does imply $Y = 0$ and $Z = 0$. Consider, for example, Y. For $m = 1$ we immediately have $Y = (0)$. For $m = 2$ we have

$$Y = \begin{pmatrix} 0 & a \\ a & 0 \end{pmatrix}, \quad a \in \mathbb{R},$$

and from (4.7) it follows that $a = 0$ and $Y = 0$. For $m = 3$ we have

$$Y = \begin{pmatrix} 0 & a & b \\ a & 0 & c \\ b & c & 0 \end{pmatrix}, \quad a, b, c \in \mathbb{R},$$

and from (4.7) it follows that $a + b = a + c = b + c = 0$, i.e. $a = b = c = 0$ and $Y = 0$. One considers Z similarly. Thus, we have shown that for $m, n \leq 3$ the matrix M also satisfies (M2), so $M \in \mathrm{CdV}(K_{m,n})$ and

$$\mu(K_{m,n}) \geq \dim \ker M = m + n - 2 \text{ for } m, n \leq 3.$$

We have seen (Corollary 3.3.9) that the largest eigenvalue of a weighted Laplace matrix on a connected bipartite graph is always simple. It follows that for $m + n \geq 3$ there is no CdV matrix M for $K_{m,n}$ with $\dim \ker M = m + n - 1$ (the existence of such a matrix would mean that 0 is the largest eigenvalue of M, but it must be simple). For $m + n = 2$ we have $m = n = 1$ and $K_{1,1} = K_2$: This was already examined under (a). Thus, we have shown that

$$\mu(K_{m,n}) = m + n - 2 \text{ for all } m, n \leq 3.$$

In particular, we have $\mu(K_5) = \mu(K_{3,3}) = 4$, which will be important for the following constructions.

The most important property of the CdV invariant is as follows:

Theorem 4.4.8 (Minor Monotonicity) *If G' is a minor of G, then $\mu(G') \leq \mu(G)$.*

The proof is given in Sect. 4.8, for which the geometric meaning of condition (M2) is clarified in Sect. 4.7 as a preparation: Very informally, (M2) states that the dimension of $\ker M$ remains controlled under small perturbations of the coefficients.

However, we first want to see for which questions the CdV invariant and the minor monotonicity can be used.

The exercises for this section are based on the article [108].

Exercise 4.4.1 For $n \geq 2$, let $O_n = (\{1, \ldots, n\}, \varnothing)$ be the empty graph with n vertices. We want to show $\mu(O_n) = 1$. Proceed as follows:

1. $\mathrm{CdV}(O_n)$ consists only of diagonal matrices.
2. A diagonal matrix is in $\mathrm{CdV}(O_n)$ if and only if exactly one diagonal entry is negative, at most one is zero, and all others are positive.
3. Show: $\mu(O_n) = 1$.
4. Deduce that for every graph G with at least two vertices one has $\mu(G) \geq 1$.

Exercise 4.4.2 Let W_n be the path graph with n vertices, $n \geq 2$ (see Exercise 1.6.4). Show:

1. All eigenvalues of weighted Laplace matrices on W_n are simple.
2. $\mu(W_n) = 1$.

Exercise 4.4.3 Let $G = (X, E)$ be a graph with $|E| \geq 1$. Let $G_1, \ldots, G_m$ be the connected components of G. We want to show:

$$\mu(G) = \max_{i \in \{1,\ldots,m\}} \mu(G_i). \tag{4.8}$$

1. Why is the condition $|E| \geq 1$ important?
2. Show " $\geq$ " in (4.8).
   Let $M \in \mathrm{CdV}(G)$ with $\mu(G) = \dim \ker M$.
3. Show that M is a block matrix, where the diagonal blocks M_i are weighted Laplace matrices for G_i.
4. Suppose there exist $i \neq j$ with $\dim \ker M_i \geq 1$ and $\dim \ker M_j \geq 1$. Choose arbitrary $0 \neq \tilde{f} \in \ker M_i$ and $0 \neq \tilde{g} \in \ker M_j$ and denote by f, g the respective extensions to X by 0. Consider the matrix $R = fg^* + gf^*$.
   (a) Show: $R \neq 0$, but $MR = 0$.
       Hint: Calculate Rg.
   (b) Find a contradiction to condition (M2) from definition 4.4.1.
   Let $i \in \{1, \ldots, m\}$ be the only index with $\dim \ker M_i \geq 1$.
5. Show that $\dim \ker M_i = \mu(G)$.
   Warning: At this point we do not yet know whether M_i is a CdV matrix for G_i.

6. Case 1: All eigenvalues of M_i are nonnegative.
   Show:
   (a) It holds $\dim \ker M_i = 1$ and $\mu(G) = 1$.
   (b) There exists a $j \in \{1, \ldots, m\}$ with $\mu(G_j) \geq 1$.
   Deduce that " $\leq$ " in (4.8) holds.
7. Case 2: M_i has negative eigenvalues.
   Show:
   (a) $M_i \in \mathrm{CdV}(G_i)$.
   (b) $\mu(G_i) \geq \mu(G)$.
   Deduce also in this case that " $\leq$ " in (4.8) holds.

Exercise 4.4.4 Use Exercises 4.4.1–4.4.3 to prove the following statement:
   A graph G satisfies $\mu(G) \leq 1$ if and only if G is a disjoint union of paths and isolated vertices.

4.5 Another Characterization of Planarity

In this section, we prove the following astonishing theorem:

Theorem 4.5.1 *A graph G is planar if and only if $\mu(G) \leq 3$.*

The theorem was discovered by Colin de Verdière [30]. Van der Holst [106] later found a more compact proof. We start with a lemma.

Lemma 4.5.2 *Let $G = (X, E)$ be a 3-connected planar graph and let M be a weighted Laplace matrix on G. Then $\dim \ker \left(M - \lambda_2(M)I\right) \leq 3$.*

Proof Proof by contradiction. Let M be a weighted Laplace matrix for G with $\dim \ker \left(M - \lambda_2(M)I\right) \geq 4$. By shifting by a constant, we can assume w.l.o.g. that $\lambda_2(M) = 0$, so $\dim \ker M \geq 4$.
   We consider an arbitrary crossing-free drawing of G. All faces are surrounded by cycles (Lemma 4.1.1), in particular, each face has at least 3 boundary vertices. Let u_1, u_2, u_3 be three distinct boundary vertices of some face F. For dimensional reasons, the linear mapping

$$\ker M \ni f \mapsto \begin{pmatrix} f(u_1) \\ f(u_2) \\ f(u_3) \end{pmatrix} \in \mathbb{R}^3$$

is not injective, so $U := \left\{ f \in \ker M : f(u_1) = f(u_2) = f(u_3) = 0 \right\}$ is a subspace of $\mathbb{R}^X$ with $\dim U \geq 1$. We choose an $f \in U \setminus \{0\}$ such that $|\operatorname{supp} f|$ is minimal. Then f has minimal support, and f has exactly two strong nodal domains by Proposition 3.3.7, which means that the sets

$$X_+ := \{x \in X : f(x) > 0\}, \quad X_- := \{x \in X : f(x) < 0\}$$

are connected in G.

Now we extend the graph G by a new vertex s, which is connected to the u_j. We draw the vertex s and the edges to the u_j within the face F. Then the new graph, which we denote by G', is again 3-connected (Exercise 1.2.3) and obviously planar.

Let now $p \in \operatorname{supp} f \equiv X_+ \cup X_-$. Thanks to Menger's theorem (Theorem 1.2.2) there are three internally vertex-disjoint paths P_j from s to p in G' and we can assume w.l.o.g. that $u_j \in P_j$ for all $j \in \{1, 2, 3\}$. Let v_j be the last vertex of P_j with $v_j \notin \operatorname{supp} f$ such that all vertices on P_j between s and v_j are also not in $\operatorname{supp} f$ (see Fig. 4.24). Then, each v_j has at least one neighbor in $\operatorname{supp} f$ (i.e., the next vertex on P_j), and $f(v_j) = 0$. From $f \in \ker M$, we obtain for each $j \in \{1, 2, 3\}$

$$0 = (Mf)(v_j) = \sum_{x \sim v_j} M_{v_j, x} f(x). \tag{4.9}$$

Note that (i) $M_{v_j, x} < 0$ for all $x \sim v_j$ and (ii) among the vertices x with $x \sim v_j$, there is at least one with $f(x) \neq 0$ (i.e., the neighbor of v_j in $\operatorname{supp} f$ constructed above). From (4.9), it follows that there is another vertex $y \sim v_j$ such that $f(x)$ and $f(y)$ have opposite signs. Thus, we have shown that each v_j has neighbors in both X_+ and X_-, see Fig. 4.25.

Now we contract all edges in X_+ and all edges in X_-. Since X_+ and X_- are connected, a new vertex x_+ and a new vertex x_- are created from X_+ and X_-, respectively, where x_+ and x_- are adjacent to all v_j. As a result, the new graph contains a subdivision of $K_{3,3}$, which includes the vertices s, $x_\pm$, and v_j, see Fig. 4.26. Then, $K_{3,3}$ is a minor of the planar graph G': Impossible according to Wagner's theorem (Theorem 4.2.3). $\qquad\square$

Fig. 4.24 Construction of s, p, u_j, v_j

Fig. 4.25 The neighbors of v_j in X_+ and X_-

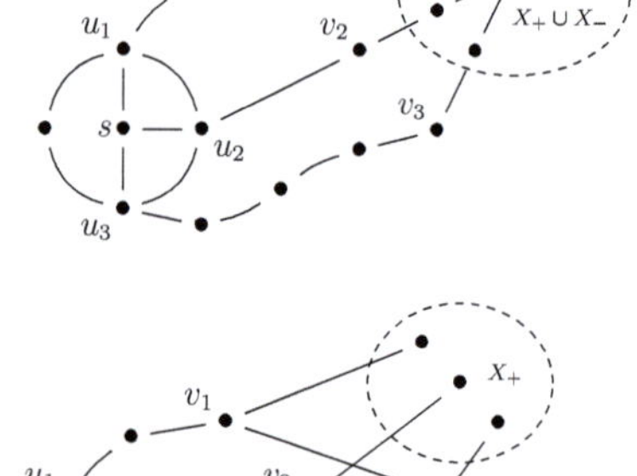

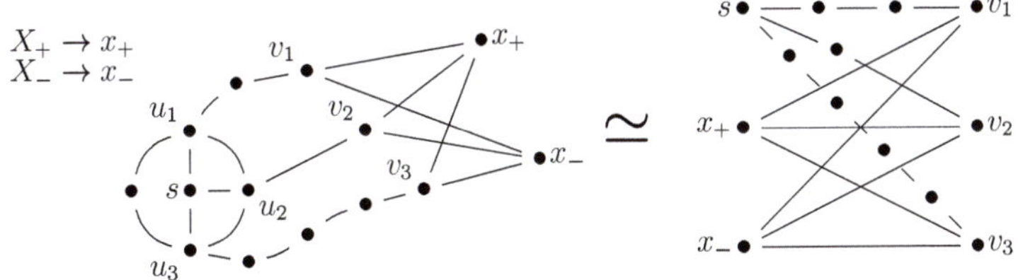

Fig. 4.26 Contraction of all edges in $X_\pm$ and the resulting subdivision of $K_{3,3}$

Proof of Theorem 4.5.1 Let G be nonplanar. According to Wagner's theorem (Theorem 4.2.3), G has a minor $G' \in \{K_5, K_{3,3}\}$. In Example 4.4.7 we showed that $\mu(G') = 4$, and from the minor monotonicity (Theorem 4.4.8) it follows that $\mu(G) \geq \mu(G') \equiv 4$.

Let G be planar. If G contains at most 3 vertices, G is a minor of K_3, so $\mu(G) \leq \mu(K_3) = 2 \leq 3$ by minor monotonicity. Therefore, we can now assume that G has at least 4 vertices. By adding additional edges, G can be extended to a maximal planar graph $\widetilde{G}$: Take any planar drawing of G and draw additional diagonals in all faces that are not triangles until a triangle graph is created (Lemma 4.3.2). Let $M \in \mathrm{CdV}(\widetilde{G})$, then $\lambda_1(M) < 0$ and $\lambda_2(M) \geq 0$ by the definition of $\mathrm{CdV}(\widetilde{G})$. For $\lambda_2(M) > 0$, it is clear that $\dim \ker M = 0 \leq 3$. For $\lambda_2(M) = 0$, one can use Lemma 4.5.2 (since $\widetilde{G}$ is a 3-connected graph according to Theorem 4.3.3) and we obtain $\dim \ker M \leq 3$. Thus $\dim \ker M \leq 3$ for all $M \in \mathrm{CdV}(\widetilde{G})$ and, therefore, $\mu(\widetilde{G}) \leq 3$. Since G is a minor of $\widetilde{G}$ by construction, it follows that $\mu(G) \leq \mu(\widetilde{G}) \leq 3$. □

The conducted study of planar graphs is a good illustration for the following very general abstract construction of graph theory [94], which we state here without proof:

Theorem 4.5.3 (Robertson-Seymour Minor Theorem) *Let $\mathcal{F}$ be an infinite countable family of graphs which is closed with respect to the formation of minors: If $G \in \mathcal{F}$, then all minors of G are also in $\mathcal{F}$. Then there are finitely many graphs $H_1, \ldots, H_n$ with the following property:*

A graph G is in $\mathcal{F}$ if and only if none of the H_j is a minor of G.

*The graphs H_j are therefore referred to as the **forbidden minors** for $\mathcal{F}$.*

Remark 4.5.4 From our constructions it follows:

- For each $m \in \mathbb{N} \cup \{0\}$, the family $\mathcal{F}_m := \{G : \mu(G) \leq m\}$ is closed with respect to the formation of minors (due to the minor monotonicity), therefore, it is uniquely determined by a finite family of forbidden minors.

- For the family $\mathcal{P}$ of all planar graphs, K_5 and $K_{3,3}$ are the forbidden minors (Wagner's theorem 4.2.3). We have just seen that $\mathcal{P} = \mathcal{F}_3$.

Some other $\mathcal{F}_m$ can also be characterized:

- by Exercise 4.4.1 it follows that $\mathcal{F}_0 = \{G = (X, E) : |X| = 1, \ E = \varnothing\}$,
- the family $\mathcal{F}_1$ consists of the linear forests (a linear forest is a disjoint union of isolated vertices and paths), see Exercise 4.4.4. Alternatively, $\mathcal{F}_1$ can be described as the set of graphs that can be drawn in $\mathbb{R}$ without crossings.
- the family $\mathcal{F}_2$ is the set of outerplanar graphs (= the planar graphs that can be drawn in the plane without crossings so that all vertices are on the boundary of the unbounded face: see Exercises 4.3.1 and 4.5.1).
- the family $\mathcal{F}_4$ is the set of graphs that allow for crossing-free linkless embeddings in $\mathbb{R}^3$ (= no two cycles in the graph are connected like links in a chain), see [78].

The structure of the families $\mathcal{F}_m$ with $m \geq 5$ is still unknown.

Remark 4.5.5 There are many other minor monotone graph invariants, see [96].

Remark 4.5.6 The **Hadwiger value** $H(G)$ of a graph G is defined by

$$H(G) := \max\left\{n \in \mathbb{N} : \ K_n \text{ is a minor of } G\right\}.$$

The famous **Hadwiger conjecture** formulated in 1943 claims that

$$\chi(G) \leq H(G) \text{ for all graphs } G.$$

It is known that the inequality holds for all graphs G with $H(G) \leq 5$. Thanks to the minor monotonicity, for all G the inequality

$$\mu(G) \geq \mu(K_{H(G)}) = H(G) - 1,$$

holds, therefore Colin de Verdière [30] formulated a weaker conjecture:

$$\chi(G) \leq \mu(G) + 1.$$

It is known that the last inequality is really fulfilled for all G with $\mu(G) \leq 4$. A general proof would also provide a new proof of the Four Color Theorem (see Remark 4.1.6).

A good introduction to the conjectures of Hadwiger and Colin de Verdière can be found in the review article by Aigner [1]. Further results about the Colin de Verdière invariant and its applications can be found in the book by Colin de Verdière [31] and in the review article by van der Holst, Lovász, Schrijver [108].

Exercise 4.5.1 Here the results of Exercise 4.3.1 are actively used.

1. Let $G = (X, E)$ be a maximal outerplanar graph and M a weighted Laplace matrix for G. We want to show:

$$\dim \ker M \leq 2. \tag{4.10}$$

   (a) Show (4.10) for $\lambda_1(M) \geq 0$.
       From now on, we assume that $\lambda_1 := \lambda_1(M) < 0$. Let $h > 0$ be the normalized eigenvector of M corresponding to λ_1 (written as a column).
   (b) Show that λ_1 is a simple eigenvalue and that h really exists and is uniquely determined.
       Consider the graph

$$\widetilde{G} := G \cup \text{ a new vertex connected to all other vertices}$$

   and the block matrix

$$\widetilde{M} := \begin{pmatrix} M & \lambda_1 h \\ \lambda_1 h^* & \lambda_1 \end{pmatrix}.$$

   Show:
   (c) $\widetilde{M}$ is a weighted Laplace matrix for $\widetilde{G}$.
   (d) $\dim \ker \widetilde{M} \geq \dim \ker M + 1$.
   (e) The inequality (4.10) is satisfied.
2. Show: A graph G satisfies $\mu(G) \leq 2$ if and only if it is outerplanar.

Exercise 4.5.2 Show with an example that the chromatic number is not minor-monotone.

4.6 Basic Concepts About Submanifolds

The remaining Sects. 4.6–4.8 are exclusively dedicated to the proof of minor monotonicity (Theorem 4.4.8). This proof requires significantly more analytic tools than all previous constructions and uses the technique of submanifolds. We will briefly repeat the corresponding notions in a minimalist form in this section. For a detailed discussion, see e.g. the books by Hirsch [64] or Lee [74].

Let U and V be finite-dimensional real vector spaces. Choose a basis $u_1, \ldots, u_q$ in U and a basis $v_1, \ldots, v_p$ in V. The expansion according to the bases identifies U with $\mathbb{R}^q$ and V with $\mathbb{R}^p$, which gives rise to a topology as well as to the concepts of a continuous or smooth function: A function $g : U \to V$ is continuous or smooth, if

$$g\left(\sum_{i=1}^{q} x_i u_i\right) = \sum_{j=1}^{p} g_j(x_1, \ldots, x_q)\, v_j, \quad (x_1, \ldots, x_q) \in \mathbb{R}^q, \tag{4.11}$$

with continuous or smooth ($= C^\infty$) functions g_j. From known results of analysis and linear algebra (transition matrices, chain rule) it follows that these properties are independent of the choice of the two bases.

If $g : U \to V$ is smooth and $u \in U$, its **differential** $dg|_u$ at the point u is the linear mapping

$$dg|_u : U \to V, \quad w \mapsto \frac{d}{dt}\Big|_{t=0} g(u + tw).$$

If g is linear, then $dg|_u = g$ for all $u \in U$. If

$$u = x_1 u_1 + \cdots + x_q u_q, \quad w = z_1 u_1 + \cdots + z_q u_q,$$

and g is given by (4.11), then

$$dg|_u(w) = \sum_{j=1}^{p} \sum_{i=1}^{q} \partial_i g_j(x_1, \ldots, x_q) z_i v_j.$$

If $\mathcal{L}(U, V)$ is the vector space of linear mappings from U to V, then from the last formula it follows that $U \ni u \mapsto dg|_u \in \mathcal{L}(U, V)$ is also smooth. If W is another finite-dimensional vector space and $h : V \to W$ is smooth, then $h \circ g : U \to W$ is also smooth and for each $u \in U$ the chain rule applies

$$d(h \circ g)|_u = dh|_{g(u)} \circ dg|_u.$$

A set $\mathcal{A} \subset V$ is a **submanifold** of dimension $p - r$ (or codimension r), if for every point $a \in \mathcal{A}$ one can find an open neighborhood Ω of a in V, a r-dimensional vector space V' and a smooth function $g : \Omega \to V'$ such that

$$\Omega \cap \mathcal{A} = \{x \in \Omega : g(x) = 0\}$$
$$\underline{and} \quad dg|_x : V \to V' \text{ is surjective for all } x \in \Omega. \tag{4.12}$$

One can always take $V' = \mathbb{R}^r$: If the above representation holds, consider an arbitrary bijective linear mapping $\Theta : V' \to \mathbb{R}^r$ and the function $\widetilde{g} := \Theta \circ g$, then $\Omega \cap \mathcal{A} = \{x \in \Omega : \widetilde{g}(x) = 0\}$ and from the bijectivity of Θ and

$$d\widetilde{g}|_x = d\Theta|_{g(x)} \circ dg|_x = \Theta \circ dg|_x$$

the surjectivity of $\widetilde{dg}|_x : V \to \mathbb{R}^r$ for all $x \in \Omega$ follows.

Let $\mathcal{A}$ be a submanifold in V and $a \in \mathcal{A}$. A **smooth curve** on $\mathcal{A}$ is a smooth mapping $\gamma : \mathcal{I} \to V$ with $\gamma(\mathcal{I}) \subset \mathcal{A}$, where $\mathcal{I} \subset \mathbb{R}$ is an open interval. The **tangent space** to $\mathcal{A}$ at the point a is defined by

$$T_a\mathcal{A} := \big\{\gamma'(0) : \gamma \text{ is a smooth curve on } \mathcal{A} \text{ with } 0 \in \mathcal{I} \text{ and } \gamma(0) = a\big\}.$$

This is a subspace of V. If $\mathcal{A}$ is represented in a neighborhood of a as in (4.12), then

$$T_a\mathcal{A} = \ker dg|_a. \tag{4.13}$$

If $\mathcal{A} \subset V$ is a subspace, then $\mathcal{A}$ is also a submanifold and for all $a \in \mathcal{A}$ we have $T_a\mathcal{A} = \mathcal{A}$.

Definition 4.6.1 Let $\mathcal{A}$ and $\mathcal{A}'$ be submanifolds in V and $a \in \mathcal{A} \cap \mathcal{A}'$. We say that $\mathcal{A}$ and $\mathcal{A}'$ **intersect transversally** at a, if

$$T_a\mathcal{A} + T_a\mathcal{A}' = V.$$

Remark 4.6.2 If V is equipped with a scalar product, the identity $(T_a\mathcal{A}+T_a\mathcal{A}')^\perp = (T_a\mathcal{A})^\perp \cap (T_a\mathcal{A}')^\perp$ holds. Therefore, the transversality condition in Definition 4.6.1 can equivalently be rewritten as

$$(T_a\mathcal{A})^\perp \cap (T_a\mathcal{A}')^\perp = \{0\}$$

In the last Sect. 4.8, the following generalization of transversality will also play a central role.

Definition 4.6.3 Let:

- U and V be finite-dimensional real vector spaces,
- $\Phi : U \to V$ be a smooth map,
- $\mathcal{S} \subset V$ be a submanifold,
- $a \in U$.

We write $\Phi \pitchfork_a \mathcal{S}$, if $\Phi(a) \in \mathcal{S}$ <u>and</u> $T_{\Phi(a)}\mathcal{S} + d\Phi|_a(U) = V$.

Remark 4.6.4 If U is a subspace (and thus a submanifold) in V, $\Phi : U \ni u \mapsto u \in V$ is the embedding and $a \in U \cap \mathcal{S}$, then $d\Phi|_a = \Phi$ and then $d\Phi|_a(U) = U \equiv T_a U$. In this case, $\Phi \pitchfork_a \mathcal{S}$ means that U and $\mathcal{S}$ intersect transversally at the point a.

Lemma 4.6.5 *Let us continue to use the notation in Definition 4.6.3. Assume $\Phi \pitchfork_a \mathcal{S}$. Then there exists an open neighborhood Ω of a in U, such that $\mathcal{A} := \Omega \cap \Phi^{-1}(\mathcal{S})$ is a submanifold in U, with $\Phi \pitchfork_x \mathcal{S}$ and $T_x\mathcal{A} = (d\Phi|_x)^{-1}(T_{\Phi(x)}\mathcal{S})$ for all $x \in \mathcal{A}$.*

Proof Since $\mathcal{S}$ is a submanifold in V, there exists an open neighborhood Ω_0 of $\Phi(a)$ in V and a smooth function $g : \Omega_0 \to \mathbb{R}^r$, such that

$$\mathcal{S} \cap \Omega_0 = \{v \in \Omega_0 : g(v) = 0\} \text{ and } dg|_v(V) = \mathbb{R}^r \text{ for all } v \in \Omega_0.$$

Since Φ is continuous, there exists an open neighborhood Ω of a in U with $\Phi(\Omega) \subset \Omega_0$, then

$$\mathcal{A} := \Omega \cap \Phi^{-1}(\mathcal{S}) \equiv \{x \in \Omega : \widetilde{g}(x) = 0\} \text{ with } \widetilde{g} := g \circ \Phi : \Omega \to \mathbb{R}^r. \tag{4.14}$$

By assumption $T_{\Phi(a)}\mathcal{S} + d\Phi|_a(U) = V$, where $T_{\Phi(a)}\mathcal{S} = \ker dg|_{\Phi(a)}$, so

$$\mathbb{R}^r = dg|_{\Phi(a)}(V) = dg|_{\Phi(a)}\big(\ker dg|_{\Phi(a)} + d\Phi|_a(U)\big) = dg|_{\Phi(a)}\big(d\Phi|_a(U)\big),$$

and thanks to the chain rule we get

$$d\widetilde{g}|_a(U) = dg|_{\Phi(a)}\big(d\Phi|_a(U)\big) = \mathbb{R}^r. \tag{4.15}$$

Choose any basis $u_1, \ldots, u_q$ in U, then

$$d\widetilde{g}|_x(U) = \text{span}\{h_1(x), \ldots, h_q(x)\} \text{ for } h_j : \Omega \ni x \mapsto d\widetilde{g}|_x(u_j) \in \mathbb{R}^r. \tag{4.16}$$

From the smoothness of $\Omega \ni x \mapsto d\widetilde{g}|_x \in \mathcal{L}(U, \mathbb{R}^r)$ it follows that all h_j are smooth, in particular continuous. Thanks to (4.15), there are distinct $m_1, \ldots, m_r \in \{1, \ldots, q\}$ such that the vectors $h_{m_1}(a), \ldots, h_{m_r}(a)$ are linearly independent. Consider the corresponding Gram matrix

$$G : \Omega \ni x \mapsto \Big(\langle h_{m_j}(x), h_{m_k}(x)\rangle\Big)_{j,k \in \{1,\ldots,r\}},$$

then the condition $\det G(x) \neq 0$ is equivalent to the linear independence of $h_{m_1}(x), \ldots, h_{m_r}(x)$. Thanks to the above choice of $m_1, \ldots, m_r$, we have $\det G(a) \neq 0$. Since all h_j are continuous, the function $x \mapsto \det G$ is also continuous, so we can assume Ω to be sufficiently small such that $\det G(x) \neq 0$ for all $x \in \Omega$. Then $h_{m_1}(x), \ldots, h_{m_r}(x)$ are linearly independent for all $x \in \Omega$, and from (4.16) it follows that $d\widetilde{g}|_x(U) = \mathbb{R}^r$ for all $x \in \Omega$. The representation (4.14) states that $\mathcal{A}$ is a submanifold.

Let $x \in \mathcal{A}$. Thanks to $d\widetilde{g}|_x = dg|_{\Phi(x)} \circ d\Phi|_x$ we obtain

$$T_x\mathcal{A} = \ker d\widetilde{g}|_x = (d\Phi|_x)^{-1}(\ker dg|_{\Phi(x)}) \equiv (d\Phi|_x)^{-1}(T_{\Phi(x)}\mathcal{S}).$$

From $\mathbb{R}^r = d\widetilde{g}|_x(U) \equiv dg|_{\Phi(x)}\big(d\Phi|_x(U)\big) \subset dg|_{\Phi(x)}(V) \subset \mathbb{R}^r$ it follows that

$$dg|_{\Phi(x)}\big(d\Phi|_x(U)\big) = dg|_{\Phi(x)}(V) = \mathbb{R}^r.$$

Thus, for each $v \in V$ there exists some $u \in U$ such that $dg|_{\Phi(x)}\big(d\Phi|_x(u)\big) = dg|_{\Phi(x)}(v)$, then $dg|_{\Phi(x)}\big(d\Phi|_x(u) - v\big) = 0$ and thus $d\Phi|_x(u) - v \in \ker dg|_{\Phi(x)}$. Therefore, the representation

$$\ker dg|_{\Phi(x)} + d\Phi|_x(U) = V$$

is proven, and due to $\ker dg|_{\Phi(x)} = T_{\Phi(x)}\mathcal{S}$ it follows that $\Phi \pitchfork_x \mathcal{S}$. $\square$

4.7 Strong Arnold Property

The condition (M2) in Definition 4.4.1, often referred to as the **strong Arnold property**, has a geometric meaning, which we explain in this section.

For $n \in \mathbb{N}$, let $\mathcal{S}_n$ be the vector space of real symmetric $n \times n$ matrices, which we endow with the following scalar product:

$$\langle M, N \rangle_{\mathcal{S}_n} := \operatorname{tr}(MN) \equiv \sum_{i,j=1}^n M_{i,j} N_{i,j} \text{ for } M, N \in \mathcal{S}_n. \tag{4.17}$$

Since every matrix $M = (M_{i,j}) \in \mathcal{S}_n$ is uniquely determined by its coefficients $M_{i,j}$ with $1 \leq i \leq j \leq n$, we have

$$\dim \mathcal{S}_n = \frac{n(n+1)}{2}.$$

Lemma 4.7.1 *For each $k \in \{0, \ldots, n\}$, the set*

$$\mathcal{S}_{n,k} := \big\{ M \in \mathcal{S}_n : \dim \ker M = k \big\}$$

is a submanifold of $\mathcal{S}_n$ of codimension $\frac{k(k+1)}{2}$. For each $M \in \mathcal{S}_{n,k}$, the representations

$$T_M \mathcal{S}_{n,k} = \big\{ N \in \mathcal{S}_n : \langle f, Nf \rangle = 0 \text{ for all } f \in \ker M \big\}, \tag{4.18}$$

$$(T_M \mathcal{S}_{n,k})^{\perp} = \{ R \in \mathcal{S}_n : MR = 0 \}. \tag{4.19}$$

hold.

Proof

(a) First consider the case $k = 0$. The function $\varphi : \mathcal{S}_n \ni M \mapsto \det M \in \mathbb{R}$ is continuous, thus $\mathcal{S}_{n,0} = \varphi^{-1}\big(\mathbb{R} \setminus \{0\}\big)$ is an open subset of $\mathcal{S}_n$ and therefore a submanifold of codimension 0. Let $M \in \mathcal{S}_{n,0}$, then $T_M \mathcal{S}_n = \mathcal{S}_n$ (due to dimension reasons) and therefore $(T_M \mathcal{S}_{n,0})^{\perp} = \{0\}$. The matrix M is invertible, so $\ker M = \{0\}$, and the sets on the right-hand sides of (4.18) and (4.19) coincide with $\mathcal{S}_n$ and $\{0\}$, respectively.

From now on, let $k \geq 1$.

(b) Let $M \in \mathcal{S}_{n,k}$. The characteristic polynomial P_M of M has the form

$$P_M(t) = \sum_{j=k}^{n} B_{n-j} t^j, \quad B_{n-k} \neq 0.$$

Since B_{n-k} is, up to sign, the sum of the principal minors of M of order $n-k$, M has a principal minor of order $n-k$ that is not zero. W.l.o.g., we will assume that the principal minor on the first $n-k$ rows and the first $n-k$ columns is not zero (otherwise one simply needs to adjust the numbering), then M is the block matrix

$$M = \begin{pmatrix} M_0 & M_1 \\ M_1^* & M_2 \end{pmatrix}$$

with an invertible real symmetric $(n-k) \times (n-k)$ matrix M_0, a real $(n-k) \times k$ matrix M_1 and a real symmetric $k \times k$ matrix M_2. By continuity, all matrices $N \in \mathcal{S}_n$ in a small open neighborhood Ω of M have the analogous representation

$$N = \begin{pmatrix} N_0 & N_1 \\ N_1^* & N_2 \end{pmatrix},$$

with invertible N_0. We have

$$N \begin{pmatrix} I_{n-k} & -N_0^{-1} N_1 \\ 0 & I_k \end{pmatrix} = \begin{pmatrix} N_0 & 0 \\ N_1^* & N_2 - N_1^* N_0^{-1} N_1. \end{pmatrix}.$$

Since the second matrix on the left-hand side is invertible, $\dim \ker N = k$ (i.e., $N \in \mathcal{S}_{n,k}$) holds if and only if $N_2 - N_1^* N_0^{-1} N_1 = 0$. Thus,

$$\mathcal{S}_{n,k} \cap \Omega = \{N \in \Omega : g(N) = 0\} \text{ with } g : \Omega \ni N \mapsto N_2 - N_1^* N_0^{-1} N_1 \in \mathcal{S}_k.$$

The function g is clearly smooth, and it remains to show that $dg|_N : \mathcal{S}_n \to \mathcal{S}_k$ is surjective for all $N \in \Omega$. Let $R \in \mathcal{S}_k$, then

$$dg|_N \left(\begin{pmatrix} 0 & 0 \\ 0 & R \end{pmatrix} \right) = \frac{d}{dt}\Big|_{t=0} g\left(N + t \begin{pmatrix} 0 & 0 \\ 0 & R \end{pmatrix} \right) = \frac{d}{dt}\Big|_{t=0} g\left(\begin{pmatrix} N_0 & N_1 \\ N_1^* & N_2 + tR \end{pmatrix} \right)$$

$$= \frac{d}{dt}\Big|_{t=0} \left((N_2 + tR) - N_1^* N_0^{-1} N_1 \right) = R,$$

so $\operatorname{ran} dg|_N = S_k$. Thus, $S_{n,k}$ is a submanifold in S_n, whose codimension coincides with $\dim S_k \equiv \frac{k(k+1)}{2}$.

(c) Consider the subspace

$$\mathcal{F} := \operatorname{span}\{ ff^* :\ f \in \ker M\} \subset S_n.$$

If $f, g \in \ker M$, then also

$$fg^* + gf^* \equiv (f+g)(f+g)^* - ff^* - gg^* \in \mathcal{F}. \tag{4.20}$$

Let $f_1, \ldots, f_k$ be any orthonormal basis in $\ker M$, then for every vector $f := a_1 f_1 + \cdots + a_k f_k \in \ker M$ (with $a_1, \ldots, a_k \in \mathbb{R}$) one has

$$ff^* = \sum_{1 \le i < j \le k} a_i a_j (f_i f_j^* + f_j f_i^*) + \sum_{i=1}^{k} a_i^2 f_i f_i^*,$$

i.e., $\mathcal{F} = \operatorname{span}\left\{ f_i f_j^* + f_j f_i^* :\ 1 \le i \le j \le k \right\}$.

We now show that the matrices $f_i f_j^* + f_j f_i^*$ are linearly independent. Let $a_{i,j} \in \mathbb{R}$ with

$$\sum_{1 \le i \le j \le k} a_{i,j}(f_i f_j^* + f_j f_i^*) = 0. \tag{4.21}$$

Choose arbitrary $1 \le \ell \le r \le k$ and multiply (4.21) from the left with f_ℓ^* and from the right with f_r, then

$$0 = \sum_{1 \le i \le j \le k} a_{i,j}(f_\ell^* f_i f_j^* f_r + f_\ell^* f_j f_i^* f_r)$$

$$\equiv \sum_{1 \le i \le j \le k} a_{i,j}\big(\langle f_\ell, f_i\rangle\langle f_j, f_r\rangle + \langle f_\ell, f_j\rangle\langle f_i, f_r\rangle\big)$$

$$= \sum_{1 \le i \le j \le k} a_{i,j}\big(\delta_{\ell,i}\delta_{j,r} + \delta_{\ell,j}\delta_{i,r}\big) = \begin{cases} a_{\ell,r}, & \ell < r, \\ 2a_{\ell,\ell}, & \ell = r, \end{cases}$$

so all $a_{i,j}$ are zero. It follows that the matrices $f_i f_j^* + f_j f_i^*$ with $1 \le i \le j \le k$ form a basis in $\mathcal{F}$, and

$$\dim \mathcal{F} = \frac{k(k+1)}{2}.$$

(d) We want to show: $T_M S_{n,k} = \mathcal{N}$ with

$$\mathcal{N} := \big\{ N \in S_n :\ \langle f, Nf\rangle = 0 \text{ for all } f \in \ker M \big\}.$$

Note that (thanks to $\operatorname{tr} AB = \operatorname{tr} BA$)

$$
\begin{aligned}
\mathcal{N} &= \left\{ N \in \mathcal{S}_n : \ \langle f, Nf \rangle = 0 \text{ for all } f \in \ker M \right\} \\
&= \left\{ N \in \mathcal{S}_n : \ \operatorname{tr}(f^* N f) = 0 \text{ for all } f \in \ker M \right\} \\
&= \left\{ N \in \mathcal{S}_n : \ \operatorname{tr}(N f f^*) = 0 \text{ for all } f \in \ker M \right\} \\
&= \left\{ N \in \mathcal{S}_n : \ \operatorname{tr}(N F) = 0 \text{ for all } F \in \mathcal{F} \right\} = \mathcal{F}^{\perp},
\end{aligned}
$$

i.e., the codimension of $\mathcal{N}$ in $\mathcal{S}_n$ is $\frac{k(k+1)}{2} = \dim \mathcal{F}$.

Let $N \in T_M \mathcal{S}_{n,k}$, then there is a smooth curve $(-\delta, \delta) \ni t \mapsto \Gamma(t) \in \mathcal{S}_{n,k} \subset \mathcal{S}_n$ with $\Gamma(0) = M$ and $\Gamma'(0) = N$. Let $P(t)$ be the matrix of orthogonal projection onto $\ker \Gamma(t)$. We want to show that for small t the mapping $t \mapsto P(t)$ is smooth. For this, we consider an orthonormal basis $e_1(t), \ldots, e_n(t)$ of eigenvectors of $\Gamma(t)$, with $\Gamma(t) = \mu_j(t) e_j(t)$ for all j, where $\mu_j(t)$ are the eigenvalues of $\Gamma(t)$. We can first assume that the μ_j are numbered in ascending order, then all μ_j are continuous (Theorem 2.2.1). Then we can renumber them so that they remain continuous and $\mu_j(t) = 0$ for $j \in \{1, \ldots, k\}$. From $\Gamma(t) \in \mathcal{S}_{n,k}$ it follows that $\mu_j(t) \neq 0$ for all $j \geq k + 1$, and then

$$
P(t) = \sum_{j=1}^{k} e_j(t) e_j(t)^*,
$$

in particular $P(t)^* = P(t)$. Since all μ_j are continuous, one can find a small $\varepsilon > 0$ and adjust δ so that $\left| \mu_j(t) \right| > \varepsilon$ for all $j \geq k + 1$ and all $t \in (-\delta, \delta)$ is fulfilled. From now on let $t \in (-\delta, \delta)$. We have

$$
\Gamma(t) = \sum_{j=1}^{n} \mu_j(t) e_j(t) e_j(t)^*
$$

and for $|z| = \varepsilon$ it holds

$$
\left(zI - \Gamma(t) \right)^{-1} = \sum_{j=1}^{n} \frac{1}{z - \mu_j(t)} \, e_j(t) e_j(t)^*.
$$

According to the Cauchy integral formula for holomorphic functions, it holds

$$
\oint_{|z|=\varepsilon} \frac{dz}{z - \mu_j(t)} =
\begin{cases}
2\pi i, & \text{if } \left| \mu_j(t) \right| < \varepsilon, \text{ i.e. for } j \in \{1, \ldots, k\}, \\
0, & \text{if } \left| \mu_j(t) \right| > \varepsilon, \text{ i.e. for } j \geq k + 1,
\end{cases}
$$

and it follows

$$P(t) = \frac{1}{2\pi i} \oint_{|z|=\varepsilon} \left(zI - \Gamma(t)\right)^{-1} dz.$$

The function on the right-hand side is smooth (Analysis: Theorem about the differentiation of integrals with parameters), which implies the smoothness of $t \mapsto P(t)$.

By construction we have $\Gamma(t)P(t) = 0$ and thus $P(t)\Gamma(t)P(t) = 0$ for all t. Since Γ and P are differentiable, one can differentiate the second identity with respect to t at the point $t = 0$:

$$0 = P'(0)\underbrace{\Gamma(0)P(0)}_{=0} + P(0)\underbrace{\Gamma'(0)}_{=N}P(0) + \underbrace{P(0)\Gamma(0)}_{=(\Gamma(0)P(0))^*=0}P'(0) = P(0)NP(0).$$

For all $f \in \ker M$ it holds $f = P(0)f$ and

$$\langle f, Nf \rangle = \langle P(0)f, NP(0)f \rangle = \langle f, \underbrace{P(0)NP(0)}_{=0} f \rangle = 0.$$

With this we have shown the inclusion $T_M \mathcal{S}_{n,k} \subset \mathcal{N}$. Since the subspaces $T_M \mathcal{S}_{n,k}$ and $\mathcal{N}$ have the same codimension $\frac{k(k+1)}{2}$, they coincide.

(e) From the previous construction it follows $(T_M \mathcal{S}_{n,k})^\perp = \mathcal{N}^\perp = (\mathcal{F}^\perp)^\perp = \mathcal{F}$. So it remains to show the equality

$$\mathcal{F} \equiv \text{span}\{ff^* : f \in \ker M\} = \{R \in \mathcal{S}_n : MR = 0\} \tag{4.22}$$

For each $f \in \ker M$ it holds $M(ff^*) = (Mf)f^* = 0f^* = 0$. Thus, from $R \in \mathcal{F}$, it follows that $MR = 0$.

Let now $R \in \mathcal{S}_n$ with $MR = 0$. Then it holds $\text{ran } R \subset \ker M$. Let $g_1, \ldots, g_r$ be an orthonormal basis in $\text{ran } R$, then $g_j \in \ker M$ and

$$R : f \mapsto \sum_{i=1}^{r} \langle h_i, f \rangle g_i$$

with suitable vectors $h_i \neq 0$. Since $R \in \mathcal{S}_n$ it holds $R^* = R$ and

$$R = R^* : f \mapsto \sum_{i=1}^{r} h_i \langle g_i, f \rangle$$

it follows $h_i \in \text{ran } R = \text{span}\{g_1, \ldots, g_r\}$, thus $h_i = a_{i,1}g_1 + \cdots + a_{i,r}g_r$ with $a_{i,j} \in \mathbb{R}$ and

$$R : f \mapsto \sum_{i,j=1}^{r} a_{i,j} \langle g_i, f \rangle g_j, \quad \text{i.e. } R = \sum_{i,j=1}^{r} a_{i,j} g_j g_i^*.$$

Since $R = R^*$ it holds $a_{i,j} = \langle g_i, R g_j \rangle = \langle R g_i, g_j \rangle = \langle g_j, R g_i \rangle = a_{j,i}$, thus

$$R = \sum_{1 \le i < j \le r} a_{i,j} (g_i g_j^* + g_j g_i^*) + \sum_{i=1}^{r} a_{i,i} g_i g_i^* \overset{(4.20)}{\in} \mathcal{F}.$$

With this we have proven (4.22).

$\square$

Let now $G = (X, E)$ be a graph with n vertices. Define

$$\mathcal{O}_G := \left\{ M \in \mathcal{S}_n : M_{x,y} = 0 \text{ for all } x \neq y \text{ with } x \not\sim y \right\}.$$

All weighted Laplace matrices for G belong to $\mathcal{O}_G$. Apparently, $\mathcal{O}_G$ is a subspace of $\mathcal{S}_n$ and thus a submanifold with $T_M \mathcal{O}_G = \mathcal{O}_G$ for all $M \in \mathcal{O}_G$. From the expression (4.17) for the scalar product in $\mathcal{S}_n$ one sees that for any $M \in \mathcal{O}_G$ the following representation holds:

$$(T_M \mathcal{O}_G)^{\perp} = (\mathcal{O}_G)^{\perp} = \left\{ R \in \mathcal{S}_n : R_{x,y} = 0 \text{ for all } x = y \text{ or } x \sim y \right\}. \tag{4.23}$$

Corollary 4.7.2 (Geometric Meaning of the Strong Arnold Property) *Let M be a weighted Laplace matrix for a graph G with exactly one negative eigenvalue. Let $k := \dim \ker M$, then the following three conditions are equivalent:*

(a) $M \in \mathrm{CdV}(G)$.
(b) $\mathcal{O}_G$ and $\mathcal{S}_{n,k}$ intersect transversally at M, i.e. $\mathcal{O}_G + T_M \mathcal{S}_{n,k} = \mathcal{S}_n$.
(c) For every matrix $S \in \mathcal{S}_n$ there is a matrix $C \in \mathcal{O}_G$, such that

$$\langle f, Sf \rangle = \langle f, Cf \rangle \text{ for all } f \in \ker M.$$

Proof The condition (M1) in Definition 4.4.1 is already fulfilled by assumption. From the representations (4.19) and (4.23) it follows that the condition (M2) exactly has the form $(T_M \mathcal{O}_G)^{\perp} \cap (T_M \mathcal{S}_{n,k})^{\perp} = \{0\}$, which means the transversality of the intersection. Thanks to (4.19), the condition in (c) is equivalent to $S - C \in T_M \mathcal{S}_{n,k}$, i.e. to $\mathcal{S}_n = \mathcal{O}_G + T_M \mathcal{S}_{n,k}$ and thus also to $\mathcal{S}_n = T_M \mathcal{O}_G + T_M \mathcal{S}_{n,k}$: This is again the condition of transversality.

$\square$

Remark 4.7.3 It can be shown that almost all (with respect to the Lebesgue measure) weighted Laplace matrices with exactly one negative eigenvalue satisfy the strong Arnold property: Essentially, this follows from the fact that for almost all such matrices the second eigenvalue is simple, which means that the strong Arnold property is fulfilled according to Lemma 4.4.6.

4.8 Proof of the Minor Monotonicity

In this section, we prove Theorem 4.4.8. The transition from a graph G to its minor G' is achieved by finitely many applications of the following graph operations:

- Removal of an isolated vertex,
- Removal of an edge,
- Contraction of an edge.

We will show that in each of these operations the CdV invariant can only decrease. The proof structure will be the same in all three cases: We take a matrix $M' \in$ CdV(G') with $\dim \ker M' = \mu(G')$ and show that we can find a matrix $M \in$ CdV(G) with $\dim \ker M \geq \dim \ker M'$. Then

$$\mu(G) \geq \dim \ker M \geq \dim \ker M' = \mu(G').$$

The existence of such a matrix M is proven by various perturbations and clever analytical constructions. The proof presented here is elaborated on the proof in van der Holst's doctoral thesis [107].

Lemma 4.8.1 (Perturbations of Transversal Intersections) *Let*

- *U, V be finite-dimensional real vector spaces,*
- *$S \subset V$ be a submanifold,*
- *$\Phi : \mathbb{R} \times U \to V$ be a smooth map,*
- *$\Phi_0 := \Phi(0, \cdot)$.*

Let $M \in U$ with $\Phi_0 \pitchfork_M S$, then for each $\varepsilon > 0$ there is a point $(a, A) \in \Phi^{-1}(S)$ with

$$\left|(a, A) - (0, M)\right| < \varepsilon, \quad a > 0, \quad \Phi \pitchfork_{(a,A)} S.$$

Proof For each $N \in U$ we have

$$d\Phi|_{(0,M)}\big((0, N)\big) = \frac{d}{dt}\Big|_{t=0} \Phi\big((0, M) + t(0, N)\big)$$

$$= \frac{d}{dt}\Big|_{t=0} \Phi_0(M + tN) = d\Phi_0|_M(N),$$

thus

$$d\Phi|_{(0,M)}(\mathbb{R} \times U) \supset d\Phi|_{(0,M)}\big(\{0\} \times U\big) = d\Phi_0|_M(U).$$

The assumption $\Phi_0 \pitchfork_M \mathcal{S}$ implies $d\Phi_0|_M(U) + T_{\Phi_0(M)}\mathcal{S} = V$, thus we get

$$d\Phi|_{(0,M)}(\mathbb{R} \times U) + T_{\Phi(0,M)}\mathcal{S} = V, \text{ i.e. } \Phi \pitchfork_{(0,M)} \mathcal{S}.$$

According to Lemma 4.6.5, there is an open neighborhood Ω of $(0, M)$ in $\mathbb{R} \times U$, such that the set $\mathcal{A} := \Phi^{-1}(\mathcal{S}) \cap \Omega$ is a submanifold, with

$$\Phi \pitchfork_{(a,A)} \mathcal{S} \text{ for all } (a, A) \in \mathcal{A}, \quad T_{(a,A)}\mathcal{A} = (d\Phi|_{(a,A)})^{-1}(T_{\Phi(a,A)}\mathcal{S}).$$

From the assumption $\Phi_0 \pitchfork_M \mathcal{S}$ it follows

$$d\Phi_0|_M(U) + T_{\Phi_0(M)}\mathcal{S} = V, \quad \text{thus} \quad d\Phi|_{(0,M)}(\{0\} \times U) + T_{\Phi(0,M)}\mathcal{S} = V.$$

In particular, there exists an $N \in U$ such that

$$\underbrace{d\Phi|_{(0,M)}(1, 0)}_{\in V} - d\Phi|_{(0,M)}(0, N) \in T_{\Phi(0,M)}\mathcal{S}.$$

It follows that $d\Phi|_{(0,M)}(1, -N) \in T_{\Phi(0,M)}\mathcal{S}$, i.e. $(1, -N) \in T_{(0,M)}\mathcal{A}$. Then one can find a smooth curve Γ on $\mathcal{A}$ with $\Gamma(0) = (0, M)$ and $\Gamma'(0) = (1, -N)$. It holds

$$\Gamma(t) = (0, M) + t(1, -N) + \mathcal{O}(t^2) \text{ for } t \to 0,$$

and one arrives at the desired result for $(a, A) := \Gamma(t)$ with sufficiently small positive t. $\qquad\square$

Lemma 4.8.2 *If G' is the graph obtained by removing an isolated vertex from G, then $\mu(G') \leq \mu(G)$.*

Proof W.l.o.g., we will assume that G has the vertices $\{1, \ldots, n\}$ and that the isolated vertex 1 is removed. Let $M' \in \mathrm{CdV}(G')$ with $\dim \ker M' = \mu(G')$. Consider

$$M := \begin{pmatrix} 1 & 0 \\ 0 & M' \end{pmatrix}.$$

Since vertex 1 in G is isolated, M is a weighted Laplace matrix for G and M (like M') has exactly one negative eigenvalue. Let $R \in \mathcal{S}_n$ such that

$$R_{i,j} = 0 \text{ for all } i = j \text{ and } i \sim j \text{ in } G$$

and $MR = 0$. From the first condition one obtains the representation

$$R = \begin{pmatrix} 0 & c \\ c^* & R' \end{pmatrix}$$

with $R' \in \mathcal{S}_{n-1}$ such that $R'_{i,j} = 0$ for all $i = j$ and $i \sim j$ in G', and from the second condition it follows that

$$0 = \begin{pmatrix} 1 & 0 \\ 0 & M' \end{pmatrix} \begin{pmatrix} 0 & c \\ c^* & R' \end{pmatrix} = \begin{pmatrix} 0 & c \\ M'c^* & M'R' \end{pmatrix},$$

thus $c = 0$ and $M'R' = 0$. Since $M' \in \mathrm{CdV}(G')$ fulfills condition (M2), it follows $R' = 0$ and thus $R = 0$. Therefore $M \in \mathrm{CdV}(G)$ and

$$\mu(M) \geq \dim \ker M = \dim \ker M' = \mu(G').$$

$$\square$$

In the following proofs, we use a special notation: For an $X \times X$ matrix $A = (A_{i,j})$ and $U, V \subset X$ we denote

$$A_{U,V} := \text{the submatrix } (A_{i,j})_{(i,j)\in U\times V},$$

$$A_{i,U} := A_{\{i\},U}, \quad A_{U,i} := A_{U,\{i\}} \text{ for } i \in X \text{ and } U \subset X.$$

Lemma 4.8.3 *If G' is the graph obtained by removing an edge from G, then $\mu(G') \leq \mu(G)$.*

Proof Let $\{1, \ldots, n\}$ be the vertices of G, and assume that G' is obtained by removing the edge $\{1, 2\}$. Denote $U := \{3, \ldots, n\}$ and consider the smooth functions

$$\Phi : \mathbb{R} \times \mathcal{O}_{G'} \to \mathcal{S}_n, \qquad \Phi(a, A) := \begin{pmatrix} A_{1,1} & -a & A_{1,U} \\ -a & A_{2,2} & A_{2,U} \\ A_{U,1} & A_{U,2} & A_{U,U} \end{pmatrix},$$

$$\Phi_0 : \mathcal{O}_{G'} \to \mathcal{S}_n, \quad \Phi_0(A) := \Phi(0, A) \equiv A.$$

Let $k := \mu(G')$, then there exists $M' \in \mathrm{CdV}(G')$ with $\dim \ker M' = k$. Then M' satisfies the strong Arnold property (M2), i.e. $\mathcal{O}_{G'} + T_{M'}\mathcal{S}_{n,k} = \mathcal{S}_n$ due to Corollary 4.7.2. Since Φ_0 is the embedding of $\mathcal{O}_{G'}$ in $\mathcal{S}_n$, it holds $\mathcal{O}_{G'} = d\Phi_0|_{M'}(\mathcal{O}_{G'})$, thus

$$d\Phi_0|_{M'}(\mathcal{O}_{G'}) + T_{M'}\mathcal{S}_{n,k} = \mathcal{S}_n, \quad \text{that is, } \Phi_0 \pitchfork_M \mathcal{S}_{n,k}.$$

Let $\varepsilon > 0$. By Lemma 4.8.1, one can find $(a, A) \in (0, \infty) \times \mathcal{O}_{G'}$ such that $\Phi(a, A) \in \mathcal{S}_{n,k}$ with $|(a, A) - (0, M')| < \varepsilon$ and $\Phi \pitchfork_{(a,A)} \mathcal{S}_{n,k}$. If ε is sufficiently small, then $M := \Phi(a, A)$ is a weighted Laplace matrix for G (since $a > 0$) with a single negative eigenvalue. Furthermore, it holds

$$\mathrm{d}\Phi|_{(a,A)}(\mathbb{R} \times \mathcal{O}_{G'}) = \Phi_{(a,A)}(\mathbb{R} \times \mathcal{O}_{G'}) = \mathcal{O}_G.$$

The condition $\Phi \pitchfork_{(a,A)} \mathcal{S}_{n,k}$ thus takes the form $\mathcal{O}_G + T_M \mathcal{S}_{n,k} = \mathcal{S}_n$, i.e., M satisfies the strong Arnold property (M2). Therefore, $M \in \mathrm{CdV}(G)$ (Corollary 4.7.2) and $\mu(G) \geq \dim \ker M = k = \mu(G')$. $\qquad\qquad\qquad\qquad\qquad\qquad\qquad\qquad\qquad\qquad\qquad\square$

Lemma 4.8.4 *Let G' be the graph obtained by contracting an edge in G, then $\mu(G') \leq \mu(G)$.*

Proof Let $\{1, \ldots, n\}$ be the vertices of G, assume that G' is obtained by contracting the edge $\{1, 2\}$, and let o be the new vertex in G'. Thus, G' has the vertices $\{o, 3, \ldots, n\}$ (in this order). Denote $U := \{3, \ldots, n\}$.

Consider the smooth functions

$$\Phi : \mathbb{R} \times \mathcal{O}_G \to \mathcal{S}_{n-1}, \quad \Phi(a, A) := \begin{pmatrix} A_{2,2} & A_{1,U} + A_{2,U} \\ A_{U,1} + A_{U,2} & A_{U,U} - aA_{U,1}A_{1,U} \end{pmatrix},$$

$$\Phi_0 : \mathcal{O}_G \to \mathcal{S}_{n-1}, \quad \Phi_0(A) := \Phi(0, A).$$

We note that Φ_0 is linear, with $\mathrm{d}\Phi_0|_A(\mathcal{O}_G) = \Phi_0(\mathcal{O}_G) = \mathcal{O}_{G'}$ for all $A \in \mathcal{O}_G$.

Let $k := \mu(G')$ and $M' \in \mathrm{CdV}(G')$ with $\dim \ker M' = k$. In particular, M' is a weighted Laplace matrix for G'. Define $A \in \mathcal{O}_G$ by

$$A := \begin{pmatrix} 0 & -1 & A_{1,U} \\ -1 & M'_{2,2} & A_{2,U} \\ A_{U,1} & A_{U,2} & M'_{U,U} \end{pmatrix},$$

with the following coefficients $A_{1,j}$ and $A_{2,j}$ for $j \in U$:

$$A_{1,j} := \begin{cases} M'_{o,j}, & \text{if } 1 \sim j \text{ and } 2 \not\sim j \text{ in } G, \\ \frac{1}{2}M'_{o,j}, & \text{if } 1 \sim j \text{ and } 2 \sim j \text{ in } G, \\ 0, & \text{if } 1 \not\sim j, \end{cases}$$

$$A_{2,j} := \begin{cases} M'_{o,j}, & \text{if } 2 \sim j \text{ and } 1 \not\sim j \text{ in } G, \\ \frac{1}{2}M'_{o,j}, & \text{if } 2 \sim j \text{ and } 1 \sim j \text{ in } G, \\ 0, & \text{if } 2 \not\sim j \text{ in } G, \end{cases}$$

then A is a weighted Laplace matrix for G with $\Phi_0(A) = M'$.

For M' we have $\mathcal{O}_{G'} + T_{M'}\mathcal{S}_{n-1,k} = \mathcal{S}_{n-1}$ (Corollary 4.7.2), and with $\mathrm{d}\Phi_0|_A(\mathcal{O}_G) = \mathcal{O}_{G'}$ and $\Phi_0(A) = M'$ we obtain

$$\mathrm{d}\Phi_0|_A(\mathcal{O}_G) + T_{\Phi_0(A)}\mathcal{S}_{n-1,k} = \mathcal{S}_{n-1}, \quad \text{i.e. } \Phi_0 \pitchfork_A \mathcal{S}_{n-1,k}.$$

Let $\varepsilon > 0$. Thanks to Lemma 4.8.1 one can find $(b, B) \in (0, \infty) \times \mathcal{O}_G$ such that

$$\Phi \pitchfork_{(b,B)} \mathcal{S}_{n-1,k} \tag{4.24}$$

and $\big|(b, B) - (0, A)\big| < \varepsilon$. We choose ε sufficiently small so that B remains a weighted Laplace matrix for G and $\Phi(b, B)$ has exactly one negative eigenvalue. Consider the $n \times n$ matrix

$$M := \begin{pmatrix} \frac{1}{b} & -\frac{1}{b} & B_{1,U} \\ -\frac{1}{b} & \frac{1}{b} + B_{2,2} & B_{2,U} \\ B_{U,1} & B_{U,2} & B_{U,U} \end{pmatrix}.$$

We want to show that $M \in \mathrm{CdV}(G)$ with $\dim \ker M = k$.

By construction, M is a weighted Laplace matrix for G, and

$$P^*MP = \begin{pmatrix} \frac{1}{b} & 0 \\ 0 & \Phi(b, B) \end{pmatrix} =: \widetilde{M} \text{ for } P := \begin{pmatrix} 1 & 1 & -bB_{1,U} \\ 0 & 1 & 0 \\ 0 & 0 & I_{n-2} \end{pmatrix}.$$

Since P is invertible, M and $\widetilde{M}$ have the same number of negative eigenvalues by Sylvester's law of inertia (Corollary 2.1.11). The number of negative eigenvalues of $\widetilde{M}$ is the same as for $\Phi(b, B)$, so M also has exactly one negative eigenvalue. In addition,

$$\dim \ker M = \dim \ker \widetilde{M} = \dim \ker \Phi(b, B) = k.$$

It remains to verify that M satisfies the strong Arnold property (M2). For this, we will use the reformulation from Corollary 4.7.2(c). Let $S \in \mathcal{S}_n$. We show that one can find a $C \in \mathcal{O}_G$ such that $\langle f, Sf \rangle = \langle f, Cf \rangle$ for all $f \in \ker M$. For this, consider the $n \times (n - 1)$ matrix

$$\Theta := \begin{pmatrix} 1 & -bB_{1,U} \\ 1 & 0 \\ 0 & I_{n-2} \end{pmatrix}$$

and denote $S' := \Theta^* S \Theta \in \mathcal{S}_{n-1}$. The condition (4.24) states

$$\mathrm{d}\Phi|_{(b,B)}(\mathbb{R} \times \mathcal{O}_G) + T_{\Phi(b,B)}\mathcal{S}_{n-1,k} = \mathcal{S}_{n-1},$$

so there exists a pair $(h, H) \in \mathbb{R} \times \mathcal{O}_G$ with $S' - \mathrm{d}\Phi|_{(b,B)}(h, H) \in T_{\Phi(b,B)}\mathcal{S}_{n-1,k}$. By Lemma 4.7.1 it holds that

$$\langle g, S'g \rangle = \big\langle g, \mathrm{d}\Phi|_{(b,B)}(h, H)g \big\rangle \text{ for all } g \in \ker \Phi(b, B). \tag{4.25}$$

We have

$$d\Phi|_{(b,B)}(h,H) = \frac{d}{dt}\Big|_{t=0}\Phi(b+th,B+tH)$$

$$= \begin{pmatrix} H_{2,2} & H_{1,U}+H_{2,U} \\ H_{U,1}+H_{U,2} & H_{U,U}-hB_{U,1}B_{1,U}-bH_{U,1}B_{1,U}-bB_{U,1}H_{1,U} \end{pmatrix}$$

$$= \Theta^* C\Theta \text{ for } C := \begin{pmatrix} \frac{h}{b^2} & -\frac{h}{b^2} & H_{1,U} \\ -\frac{h}{b^2} & H_{2,2}+\frac{h}{b^2} & H_{2,U} \\ H_{U,1} & H_{U,2} & H_{U,U} \end{pmatrix} \in \mathcal{O}_G.$$

Let $f \in \ker M$. The identity $Mf = 0$ takes the form

$$\begin{pmatrix} \frac{1}{b} & -\frac{1}{b} & B_{1,U} \\ -\frac{1}{b} & \frac{1}{b}+B_{2,2} & B_{2,U} \\ B_{U,1} & B_{U,2} & B_{U,U} \end{pmatrix}\begin{pmatrix} f_1 \\ f_2 \\ f_U \end{pmatrix} = \begin{pmatrix} 0 \\ 0 \\ 0 \end{pmatrix}.$$

From the first row we get $f_1 = f_2 - bB_{1,U}f_U$, and the substitution into the remaining two rows yields

$$0 = -\frac{1}{b}(f_2 - bB_{1,U}f_U) + \left(\frac{1}{b}+B_{2,2}\right)f_2 + B_{2,U}f_U$$

$$= B_{2,2}f_2 + (B_{1,U}+B_{2,U})f_U,$$

$$0 = B_{U,1}(f_2 - bB_{1,U}f_U) + B_{U,2}f_2 + B_{U,U}f_U$$

$$= (B_{U,1}+B_{U,2})f_2 + (B_{U,U}-bB_{U,1}B_{1,U})f_U,$$

i.e.

$$\Phi(b,B)g = 0 \text{ for } g := \begin{pmatrix} f_2 \\ f_U \end{pmatrix}, \tag{4.26}$$

where $\Theta g = f$. Thus, using (4.25) for all $f \in \ker M$ and the corresponding g from (4.26), we get:

$$\langle f, Sf\rangle = \langle \Theta g, S\Theta g\rangle = \langle g, \Theta^* S\Theta g\rangle = \langle g, S'g\rangle$$

$$= \big(g, d\Phi|_{(b,B)}(h,H)g\big) = \langle g, \Theta^* C\Theta g\rangle = \langle \Theta g, C\Theta g\rangle = \langle f, Cf\rangle.$$

Thus, we have shown $M \in \mathrm{CdV}(G)$, hence, $\mu(G) \geq \dim\ker M = k = \mu(G')$.   □

Remark 4.8.5 The invariant of Colin de Verdière also originates from the study of differential operators. Let X be a connected compact Riemannian manifold and let $m(X)$ be the maximum multiplicity of the second smallest eigenvalue of a real symmetric elliptic differential operator of second order on X. For $\dim X = 1$, it is clear that $m(X) = 2$, for $\dim X \geq 3$ one obtains $m(X) = \infty$, while for $\dim X = 2$ there are various upper bounds for $m(X)$ that take into account topological properties of X [28]. Colin de Verdière [29] showed that for a graph G that can be drawn on X without crossings, the inequality $\mu(G) \leq m(X)$ holds. For the proof, a drawing of G on X was used to "extend" a weighted Laplace matrix on G to an elliptic differential operator on X (the strong Arnold property is necessary for this construction). A relationship to planarity arose because a graph is planar if and only if it can be drawn on the two-dimensional sphere $\mathbb{S}^2$ without crossings (since drawings on the plane and on the sphere can be transformed into each other using stereographic projection) and because it was possible to explicitly calculate $m(\mathbb{S}^2) = 3$.

Remark 4.8.6 There exist further graphical representations of graphs [23]. For example, one can represent the vertices $x \in X$ of a planar graph $G = (X, E)$ by pairwise disjoint open circular disks S_x in $\mathbb{R}^2$ or $\mathbb{S}^2$ such that $x \sim y$ is equivalent to $\overline{S}_x \cap \overline{S}_y \neq \varnothing$ (Koebe-Andreev-Thurston circle packing theorem). Using this representation, Spielman and Teng [99] showed that for every planar graph $G = (X, E)$ with the Laplace matrix L, the inequality

$$\lambda_2(L) \leq \frac{8 \deg_{\max} G}{|X|}$$

is satisfied. The proof method was then extended to further classes of graphs. In particular, Amini and Cohen-Steiner [7] discovered that a universal constant $C > 0$ exists such that for all graphs G that can be drawn without crossings on a compact surface of genus g, the inequalities

$$\lambda_k(N) \leq C \frac{(g + k) \deg_{\max} G}{|X|}, \quad k \in \{1, \dots, |X|\},$$

hold for the eigenvalues of the associated normalized Laplace matrices N. Such estimates find a direct application in cluster analysis using the Cheeger inequalities (Sect. 3.1). Similar inequalities were previously found for the eigenvalues of differential operators on compact manifolds [63]. A detailed discussion of further geometric and analytical aspects of graph theory can be found in the book [76] by Lovász.

Remark 4.8.7 As a further generalization of graphs, so-called metric graphs are considered. A metric graph is obtained by viewing the edges of a graph as intervals. Such structures represent a kind of interpolation between graphs and manifolds, and various differential operators can be introduced in a canonical way, whose eigenvalues and eigenfunctions reflect important geometric properties. We refer to the books by Berkolaiko and Kuchment [16], Kostenko and Nicolussi [70], Kurasov [72], Mugnolo [85] and Post [92] for a quick introduction to this active field of research.

References

1. M. Aigner, Geometrische Darstellungen von Graphen und der 4-*Farbensatz*, in *Angewandte Mathematik, insbesondere Informatik*, Hrsg. by P. Horster (Vieweg, Wiesbaden, 1999)
2. M. Aigner, *Graphentheorie* (Springer Spektrum, Heidelberg, 2015)
3. M. Aigner, G.M. Ziegler, *Proofs from the BOOK*, 6th edn. (Springer, Berlin, 2018)
4. N. Alon, Eigenvalues and expanders. Combinatorica **6**(2), 83–96 (1986)
5. L. Alon, M. Goresky, Morse theory for discrete magnetic operators and nodal count distribution for graphs. J. Spectr. Theory **13**, 1225–1260 (2023)
6. N. Alon, V.D. Milman: λ_1, isoperimetric inequalities for graphs and superconcentrators. J. Combin. Theory Ser. B **38**, 73–88 (1985)
7. O. Amini, D. Cohen-Steiner, A transfer principle and applications to eigenvalue estimates for graphs. Comment. Math. Helv. **93**(1), 203–223 (2018)
8. K. Appel, W. Haken, Every planar map is four colorable. I. Discharging. Illinois J. Math. **21**(3) 429–490 (1977)
9. K. Appel, W. Haken, Every planar map is four colorable. II. Reducibility. Illinois J. Math. **21**(3), 491–567 (1977)
10. R. Band, The nodal count $\{0, 1, 2, 3, \dots\}$ implies the graph is a tree. Philos. Trans. R. Soc. Lond. Ser. A Math. Phys. Eng. Sci. **372**, 20120504 (2014)
11. R.B. Bapat, *Graphs and Matrices* (Springer, London, 2014)
12. M.T. Barlow, *Random Walks and Heat Kernels on Graphs* (Cambridge University Press, Cambridge, 2017)
13. F. Belardo, S.M. Cioabă, J. Koolen, J. Wang, Open problems in the spectral theory of signed graphs. Art Discr. Appl. Math. **1**(2), 2.10 (2018)
14. G. Berkolaiko, A lower bound for nodal count on discrete and metric graphs. Comm. Math. Phys. **278**(3), 803–819 (2008)
15. G. Berkolaiko, Nodal count of graph eigenfunctions via magnetic perturbation. Anal. PDE **6**(5), 1213–1233 (2013)
16. G. Berkolaiko, P. Kuchment, *Introduction to Quantum Graphs* (AMS, Providence, 2013)
17. G. Berkolaiko, J.B. Kennedy, P. Kurasov, D. Mugnolo, Edge connectivity and the spectral gap of combinatorial and quantum graphs. J. Phys. A Math. Theor. **50**, 365201 (2017)
18. N. Biggs, *Algebraic Graph Theory* (Cambridge University Press, Cambridge, 2012)
19. T. Biyikoğlu, J. Leydold, P.F. Stadler, *Laplacian Eigenvectors of Graphs*. Lecture Notes Math., vol. 1915 (Springer, Berlin, 2007)
20. T. Biyikoğlu, J. Leydold, P.F. Stadler: Nodal domain theorems and bipartite subgraphs. Electron. J. Linear Algebra **13**, 344–351 (2005)
21. B. Bollobás, *Modern Graph Theory* (Springer, New York, 1998)
22. A. Bretto, A. Faisant, F. Hennecart, *Elements of Graph Theory* (EMS Press, London, 2022)

K. Naderi, K. Pankrashkin, *Introduction to Spectral Graph Theory*, Compact Textbooks in Mathematics, https://doi.org/10.1007/978-3-032-01708-6

23. G.R. Brightwell, E.R. Scheinerman, Representations of planar graphs. SIAM J. Discrete Math. **6**(2), 214–229 (1993)
24. A.E. Brouwer, W.H. Haemers, *Spectra of Graphs* (Springer, New York, 2012)
25. J. Brüning, Nodal sets in mathematical physics. Eur. Phys. J. Special Top. **145**, 181–189 (2007)
26. J. Cheeger, A lower bound for the lowest eigenvalue of the Laplacian, in *Problems in Analysis*, ed. by R.C. Gunning (Princeton University Press, Princeton, 1970), pp. 195–199
27. F.R.K. Chung, *Spectral Graph Theory* (AMS, Providence, 1997)
28. B. Colbois, Y. Colin de Verdière, Multiplicité de la première valeur propre du laplacien des surfaces à courbure constante. Comment. Math. Helvet. **63**, 194–208 (1988)
29. Y. Colin de Verdière, Construction de laplaciens dont une partie finie du spectre est donnée. Ann. Sci. École Norm. Sup. **20**, 599–615 (1987)
30. Y. Colin de Verdière, Sur un nouvel invariant des graphes et un critère de planarité. J. Combin. Theory Ser. B **50**(1), 11–21 (1990). English translation: Y. Colin de Verdière, On a new graph invariant and a criterion for planarity, in *Graph Structure Theory: Proc. AMS–IMS–SIAM Joint Summer Research Conference on Graph Minors*, ed. by N. Robertson, P. Seymour. Contemp. Math., vol. 147 (American Mathematical Society, Providence), pp. 137–147
31. Y. Colin de Verdière, *Spectres de Graphes*. Cours spécialisés, vol. 4 (SMF, London, 1998)
32. Y. Colin de Verdière, Magnetic interpretation of the nodal defect on graphs. Anal. PDE **6**(5), 1235–1242 (2013)
33. L. Collatz, U. Sinogowitz, Spektren endlicher Graphen. Abh. Math. Seminar Univ. Hamburg **2**, 63–77 (1957)
34. R. Courant, D. Hilbert, *Methods of Mathematical Physics. Volume 1* (Wiley-VCH, Weinheim, 1989)
35. D. Cvetković, *Bihromatičnost i spektar grafa*, in *Uvođenje mladih u naučni rad*. VI., Hrsg. by R.R. Janić. Matematička Biblioteka, vol. 41 (Beograd, 1969), pp. 193–194
36. D. Cvetković, Chromatic number and the spectrum of a graph. Publ. Inst. Math. (Belgrade) **14**, 25–38 (1972)
37. D. Cvetković, S. Simić, Graph spectra in computer science. Lin. Alg. Appl. **434**, 1545–1562 (2011)
38. D. Cvetković, M. Doob, H. Sachs, *Spectra of Graphs* (Academic Press, New York, 1980)
39. D.M. Cvetković, M. Doob, I. Gutman, A. Torgašev, *Recent Results in the Theory of Graph Spectra* (North-Holland, Amsterdam, 1988)
40. D. Cvetković, P. Rowlinson, S.K. Simić, Signless Laplacians of finite graphs. Lin. Alg. Appl. **423**, 155–171 (2007)
41. D. Cvetković, P. Rowlinson, S. Simić, *An Introduction to the Theory of Graph Spectra* (Cambridge University Press, Cambridge, 2010)
42. D. Cvetković, P. Rowlinson, S. Simić, *Eigenspaces of Graphs* (Cambridge University Press, Cambridge, 2011)
43. E.B. Davies, G.M.L. Gladwell, J. Leydold, P.F. Stadler, Discrete nodal domain theorems. Linear Algebra Appl. **336**, 51–60 (2001)
44. L. de Lima, V. Nikiforov, C. Oliveira, The clique number and the smallest Q-*eigenvalue of graphs*. Discrete Math. **339**, 1744–1752 (2016)
45. R. Diestel, *Graph Theory* (Springer, Berlin, 2017)
46. J. Dodziuk, Difference equations, isoperimetric inequality and transience of certain random walks. Trans. AMS **284**(2), 787–794 (1984)
47. J. Dodziuk, W.S. Kendall, Combinatorial Laplacians and isoperimetric inequality, in *From Local Times to Global Geometry, Control and Physics*, ed. by K.D. Elworthy. Research Notes in Mathematics Series, vol. 150 (Pitman, London, 1986), pp. 68–74
48. A.M. Duval, V. Reiner, Perron-Frobenius type results and discrete versions of nodal domain theorems. Linear Algebra Appl. **294**, 259–268 (1999)
49. P. Erdős, A. Rényi, V.T. Sós, On a problem of graph theory. Stud. Sci. Math. Hungar. **1**, 215–235 (1966)
50. M. Fiedler, Algebraic connectivity of graphs. Czech. Math. J. **23**(2), 298–305 (1973)

51. M. Fiedler, Eigenvectors of acyclic matrices. Czech. Math. J. **25**, 607–618 (1975)
52. M. Fiedler, A property of eigenvectors of nonnegative symmetric matrices and its application to graph theory. Czech. Math. J. **25**(4), 619–633 (1975)
53. C. Ge, S. Liu, Symmetric matrices, signed graphs, and nodal domains. Calc. Var. PDE **62**, 137 (2023)
54. G.M.L. Gladwell, H. Zhu, Courant's nodal line theorem and its discrete counterparts. Q. J. Mech. Appl. Math. **55**(1), 1–15 (2002)
55. C.D. Godsil, M.W. Newman, Eigenvalue bounds for independent sets. J. Combin. Theory Ser. B **98**(4), 721–734 (2008)
56. C. Godsil, G. Royle, *Algebraic Graph Theory* (Springer, New York, 2001)
57. R.L. Graham, H.O. Pollak, On the addressing problem for loop switching. Bell Syst. Tech. J. **50**(8), 2495–2519 (1971)
58. R.L. Graham, H.O. Pollak, On embedding graphs in squashed cubes, in *Graph Theory and Applications*, ed. by Y. Alavi, D.R. Lick, A.T. White. Lecture Notes Math., vol. 303 (Springer, Berlin, 1972), pp. 99–110
59. D. Grieser, The first eigenvalue of the Laplacian, isoperimetric constants, and the max flow min cut theorem. Arch. Math. (Basel) **87**, 75–85 (2006)
60. A. Grigor'yan, *Introduction to Analysis on Graphs* (AMS, Providence, 2018)
61. W.H. Haemers, Interlacing eigenvalues and graphs. Lin. Alg. Appl. **227–228**, 593–616 (1995)
62. J. Harant, S. Richter, A new eigenvalue bound for independent sets. Discr. Math. **338**(10), 1763–1765 (2015)
63. A. Hassannezhad, Conformal upper bounds for the eigenvalues of the Laplacian and Steklov problem. J. Funct. Anal. **261**, 3419–3436 (2011)
64. M.W. Hirsch, *Differential Topology* (Springer, New York, 2012)
65. A.J. Hoffman, On eigenvalues and colorings of graphs, in *Graph Theory and Its Applications*, ed. by B. Harris (Academic Press, New York, 1970), pp. 93–112
66. C. Hoppen, D.P. Jacobs, V. Trevisan, *Locating Eigenvalues in Graphs*. SpringerBriefs in Mathematics (Springer, Warszawa, 2022)
67. H. Huang, Induced subgraphs of hypercubes and a proof of the sensitivity conjecture. Ann. Math. **190**, 949–955 (2019)
68. M. Keller, M. Schwarz, Courant's nodal domain theorem for positivity preserving forms. J. Spectr. Theory **10**, 271–309 (2020)
69. M. Keller, D. Lenz, R.K. Wojciechowski, *Graphs and Discrete Dirichlet Spaces* (Springer, Cham, 2021)
70. A. Kostenko, N. Nicolussi, *Laplacians on Infinite Graphs* (EMS Press, Berlin, 2023)
71. J. Kunegis, S. Schmidt, A. Lommatzsch, J. Lerner, E.W. De Luca, S. Albayrak, Spectral analysis of signed graphs for clustering, prediction and visualization, in *Proceedings of the 2010 SIAM International Conference on Data Mining* (2010). https://doi.org/10.1137/1.9781611972801.49
72. P. Kurasov, *Spectral Geometry of Graphs*. Operator Theory: Adv. Appl., vol. 293 (Birkhäuser, Basel, 2023)
73. K. Kuratowski, Sur le problème des courbes gauches en topologie. Fundam. Math. **15**, 271–283 (1930)
74. J. Lee, *Introduction to Smooth Manifolds* (Springer, New York, 2003)
75. J.R. Lee, S. Oveis Gharan, L. Trevisan, Multiway spectral partitioning and higher-order Cheeger inequalities. J. ACM **61**(6), 37 (2014)
76. L. Lovász, *Graphs and Geometry* (AMS, Providence, 2019)
77. L. Lovász, J. Pelikán, On the eigenvalues of trees. Period. Math. Hung. **3**, 175–182 (1973)
78. L. Lovász, A. Schrijver, A Borsuk theorem for antipodal links and a spectral characterization of linklessly embeddable graphs. Proc. AMS **126**(5), 1275–1285 (1998)
79. M.W. Mahoney, *Lecture Notes on Spectral Graph Methods* (2016), https://arxiv.org/abs/1608.04845
80. W. McCuaig, A simple proof of Menger's theorem. J. Graph Theory **8**, 427–429 (1984)

81. P. van Mieghem, *Graph Spectra for Complex Networks* (Cambridge University Press, Cambridge, 2011)

82. B. Mohar, Isoperimetric numbers of graphs. J. Combin. Theory Ser. B **47**, 274–291 (1989)

83. B. Mohar, W. Woess, A survey on spectra of infinite graphs. Bull. LMS **21**(3), 209–234 (1989)

84. T. Motzkin, E.G. Straus, Maxima for graphs and a new proof of a theorem of Turán. Canad. J. Math. **17**, 533–540 (1965)

85. D. Mugnolo, *Semigroup Methods for Evolution Equations on Networks* (Springer, Cham, 2014)

86. R. Mulas, D. Horak, J. Jost, Graphs, simplicial complexes and hypergraphs: spectral theory and topology, in *Higher-order Systems*, ed. by F. Battiston, G. Petri (Springer, Cham, 2022), pp. 11–58

87. B. Nica, *A Brief Introduction to Spectral Graph Theory* (EMS Press, Berlin, 2018)

88. V. Nikiforov, Some inequalities for the largest eigenvalue of a graph. Combin. Prob. Comput. **11**, 179–189 (2002)

89. I. Oren, Nodal domain counts and the chromatic number of graphs. J. Phys. A Math. Theor. **40**, 9825–9832 (2007)

90. G. Oshikiri, Cheeger constant and connectivity. Interdisc. Inform. Sci. **8**(2), 147–150 (2002)

91. Å. Pleijel, Remarks on Courant's nodal line theorem. Comm. Pure. Appl. Math. **9**(3), 543–550 (1956)

92. O. Post, *Spectral Analysis on Graph-like Spaces*. Lecture Notes Math., vol. 2039 (Springer, Berlin, 2012)

93. O. Post, Analysis on graphs, in *Lecture Notes, ICMAT* (2017). https://www.icmat.es/escuelaJAE/2017/Cursos/Post.pdf

94. N. Robertson, P. Seymour, Graph minors. XX. Wagner's conjecture. J. Combin. Theory Ser. B **92**(2), 325–357 (2004)

95. R. Roth, On the eigenvectors belonging to the minimum eigenvalue of an essentially nonnegative symmetric matrix with bipartite graph. Lin. Alg. Appl. **118**, 1–10 (1989)

96. A. Schrijver, Minor-monotone graph invariants, in *Surveys in Combinatorics, 1997*, ed. by R.A. Bailey. LMS Lecture Note Series, vol. 241 (Cambridge University Press, Cambridge, 1997), pp. 163–196

97. P.M. Soardi, *Potential Theory on Infinite Networks*. Lecture Notes in Mathematics, vol. 1590 (Springer, Berlin, 1994)

98. D.A. Spielman, Spectral and algebraic graph theory. Book draft (2025). http://cs-www.cs.yale.edu/homes/spielman/sagt

99. D.A. Spielman, S.-H. Teng, Spectral partitioning works: Planar graphs and finite element meshes. Linear Algebra Appl. **421**(2–3), 284–305 (2007)

100. Z. Stanić, *Inequalities for Graph Eigenvalues* (Cambridge University Press, Cambridge, 2015)

101. Z. Stanić, Some properties of the eigenvalues of the net Laplacian matrix of a signed graph. Discuss. Math. Graph Theory **42**, 893–903 (2022)

102. S. Steinerberger, A spectral approach to the shortest path problem. Linear Algebra Appl. **620**, 182–200 (2021)

103. D. Stevanović, *Spectral Radius of Graphs* (Academic Press, Cham, 2014)

104. T. Sunada, Discrete geometric analysis, in *Analysis on Graphs and Its Applications*, ed. by P. Exner, J.P. Keating, P. Kuchment, T. Sunada, A. Teplyaev. Proc. Symp. Pure Math., vol. 77 (AMS, Providence, 2008), pp. 51–83

105. L. Trevisan, *Lecture notes on Graph Partitioning, Expanders and Spectral Methods*. https://lucatrevisan.github.io/books/expanders-2016.pdf

106. H. van der Holst, A short proof of the planarity characterization of Colin de Verdière. J. Combin. Theory Ser. B **65**, 269–272 (1995)

107. H. van der Holst, Topological and spectral graph characterizations. Dissertation, University of Amsterdam, 1996

108. H. van der Holst, L. Lovász, A. Schrijver, The Colin der Verdière graph parameter, in *Graph Theory and Combinatorial Biology*, ed. by L. Lovász, A. Gyárfás, G. Katona, A. Recski,

L. Székely. Bolyai Soc. Math. Stud., vol. 7 (János Bolyai Math. Soc., Budapest, 1999), pp. 29–85

109. L. Volkmann, *Graphen und Digraphen. Eine Einführung in die Graphentheorie* (Springer, Berlin, 1996)

110. U. von Luxburg, A tutorial on spectral clustering. Stat. Comput. **17**, 395–416 (2007)

111. K. Wagner, Über eine Erweiterung des Satzes von Kuratowski. Deutsche Math. **2**, 280–285 (1937)

112. W. Walter, *Ordinary Differential Equations* (Springer, New York, 1998)

113. D.B. West, *Introduction to Graph Theory* (Prentice Hall, Harlow, 2001)

114. H.S. Wilf, The eigenvalues of a graph and its chromatic number. J. Lond. Math. Soc. **42**, 330–332 (1967)

115. H.S. Wilf, Spectral bounds for the clique and independence number of graphs. J. Combin. Theory Ser. B **40**, 113–117 (1986)

116. D.P. Williamson, Lecture notes of the course ORIE 6334: Spectral graph theory. Cornell University (2016 Fall). https://people.orie.cornell.edu/dpw/orie6334/Fall2016/

117. P. Wocjan, C. Elphick, New spectral bounds on the chromatic number encompassing all eigenvalues of the adjacency matrix. Electr. J. Combin. **20**(3), P39 (2013)

118. W. Woess, *Random Walks on Infinite Graphs and Groups*. Cambridge Tracts Math., vol. 138 (Cambridge University Press, Cambridge, 2000)

119. H. Xu, S.-T. Yau, Nodal domain and eigenvalue multiplicity of graphs. J. Combin. **3**(4), 609–622 (2012)

120. H. Yuan, A bound for the spectral radius of graphs. Linear Algebra Appl. **108**, 135–139 (1988)

121. S. Zelditch, Eigenfunctions and nodal sets, in *Geometry and Topology*, ed. by H.-D. Cao, S.-T. Yau. Surveys in Differential Geometry, vol. 18 (International Press, New York, 2013), pp. 237–308

Index

K. Naderi, K. Pankrashkin, *Introduction to Spectral Graph Theory*, Compact Textbooks in Mathematics, https://doi.org/10.1007/978-3-032-01708-6